Systems thinking in sport is an essential read for all of those interested in understanding complex problems in high performance. The book provides a fantastic outline of the methodological approaches that can be used to both collect and analyse data to create insights from a systems thinking perspective. Each chapter provides content that is easy to understand and use. This makes it a brilliant resource for those like me who are interested in these research approaches but have little experience in the range of techniques that can be used or how to use them. As such, the book is a "one stop shop" in its ability to provide comprehensiveness in one volume. This is definitely an essential text for those thinking a little differently about research in sport and exercise science.

Professor Barry Drust, *Director of The Graduate School of Sport and Professional Practice. School of Sport, Exercise and Rehabilitation Sciences. University of Birmingham*

Complexity and systems problems exist throughout all aspects of athlete preparation in elite sports. This book provides a brilliant practical guide to understanding the fundamental structure and cause of many of these issues. This book gave me a deeper appreciation for the true cause of problems in sport and signposted ways to solve them. This is a fascinating and highly influential book from authors with a great breadth of experience within and outside of elite sports. Having read this, I now see systems problems everywhere. Incredibly useful!

Patrick Hogben, *Head of Strength and Conditioning, Atlanta Hawks, National Basketball Association (NBA)*

SYSTEMS THINKING METHODS IN SPORT

Sport is increasingly being described as a complex system. This inherent complexity cannot be understood by examining components in isolation; rather, the system as a whole should represent the unit of analysis. Systems thinking is the answer to understanding this complexity and is gaining traction in sport. Systems thinking provides a philosophy and a set of associated methods which can be used to understand and optimise the behaviour of complex systems, such as those inherent within sport.

This book presents, for the first time, a practical guide to applying contemporary systems thinking methods in sport as well as case study applications demonstrating how their outputs can be translated in practice. The methods described in this book can be used for better understanding the systemic influences in a broad range of sport contexts, including performance, injury, team functioning, decision-making, adverse incidents, sports organisation design and redesign, technology implementation, and proactive risk assessments.

Systems Thinking Methods in Sport provides a practical step-by-step guide for sports practitioners and stakeholders, as well as university students and academics in applying state-of-the-art systems thinking methods to sport.

SYSTEMS THINKING METHODS IN SPORT

Practical Guidance and Case Study Applications

Scott McLean, Mitchell Naughton, Gemma J.M. Read, Neville A. Stanton, Adam Hulme, Guy H. Walker, and Paul M. Salmon

NEW YORK AND LONDON

Designed cover image: Getty Images

First published 2025
by Routledge
605 Third Avenue, New York, NY 10158

and by Routledge
4 Park Square, Milton Park, Abingdon, Oxon, OX14 4RN

Routledge is an imprint of the Taylor & Francis Group, an informa business

ISBN: 978-1-032-19498-1 (hbk)
ISBN: 978-1-032-19497-4 (pbk)
ISBN: 978-1-003-25947-3 (ebk)

DOI: 10.4324/9781003259473

Typeset in Sabon
by codeMantra

CONTENTS

FIGURES

TABLES

ABOUT THE AUTHORS

Scott McLean, PhD, is a Senior Research Fellow and the research theme leader for Sport and Outdoor Recreation in the Centre for Human Factor and Sociotechnical Systems at the University of the Sunshine Coast. He holds a bachelor's degree in sport and exercise science, a master's degree in exercise physiology, and a PhD in human factors and ergonomics in sport. Scott's research spans a broad range of domains, including sport science, safety science, systems thinking, and complexity science. He has strong collaborations in sport and has worked and conducted research with multiple national and international sporting organisations, including the World Anti-Doping Agency, Sport Integrity Australia, the Australian Institute of Sport, Athletics Australia, Cycling Australia, the French Anti-Doping Agency, Sport and Recreation Victoria, the English Institute of Sport, Scottish Rugby Union, the Defence Science and Technology Group, St Kilda AFL Club, Brisbane Roar Football Club, Sunshine Coast Lightning Netball, and UniSC Para Sport.

Mitchell Naughton, PhD, is an Associate Lecturer in exercise physiology at the University of Newcastle (Australia). He has undergraduate and postgraduate degrees in sport and exercise science and earned a PhD from the University of the Sunshine Coast Centre for Human Factors and Sociotechnical Systems, where he investigated the influence of external loads on post-match fatigue in collision sport athletes. He has over ten years of applied and research experience in the areas of performance analysis, sports science, and exercise physiology. He has presented at leading international and national sports science conferences and maintains an active research profile with international collaborators across a range of domains, including sport, physical activity, defence, and occupational settings.

Gemma J.M. Read, PhD, is the Director of the Centre for Human Factors and Sociotechnical Systems at the University of the Sunshine Coast. She has degrees in behavioural science and law and earned a PhD from the Monash University Accident Research Centre. She has over 16 years' experience applying human factors and systems thinking methods in both academia and government roles. Her work spans a range of domains, including transportation, healthcare, construction, defence, sport, and outdoor recreation, and her research has been recognised by awards from the UK Chartered Institute of Ergonomics and Human Factors, the US Human Factors and Ergonomics Society, and the Australian Aviation Psychology Association.

Neville A. Stanton, PhD, DSc, is a chartered psychologist, chartered ergonomist, and chartered engineer. He is Professor Emeritus of Human Factors Engineering in the Faculty of Engineering and Physical Sciences at the University of Southampton in the UK. He has degrees in occupational psychology, applied psychology, and human factors engineering, and he has worked at the Universities of Aston, Brunel, Cornell, and the Massachusetts Institute of Technology. His research interests include modelling, predicting, analysing, and evaluating human performance in systems as well as designing the interfaces and interaction between humans and technology. Professor Stanton has worked on the design of automobiles, aircraft, ships, and control rooms over the past 30 years on a variety of automation projects. He has published 60 books and over 400 journal papers on ergonomics and human factors. In 1998, he was presented with the Institution of Electrical Engineers Divisional Premium Award for research into system safety. The Institute of Ergonomics and Human Factors in the UK awarded him The Otto Edholm Medal in 2001, The President's Medal in 2008 and 2018, The Sir Frederic Bartlett Medal in 2012, and The William Floyd Medal in 2019 for his contributions to basic and applied ergonomics research. The Royal Aeronautical Society awarded him and his colleagues the Hodgson Prize in 2006 for research on design-induced, flight-deck errors published in *The Aeronautical Journal*. The University of Southampton has awarded him a Doctor of Science in 2014 for his sustained contribution to the development and validation of human factors methods.

Adam Hulme, PhD, is an Australian Research Council DECRA Fellow specialising in the use of systems science methods at the University of Queensland, Brisbane, Australia. He holds a BSc in sport and exercise science, an honours in sports psychology, a master's in health promotion, and a PhD in epidemiology and systems human factors in the area of sports injury aetiology. His doctoral programme was completed at an International Olympic Committee (IOC) world-leading research centre in sports injury prevention. He has an extensive list of research publications in sport, including a world-first complex systems microsimulation. In addition to sport, Dr Hulme has applied and evaluated systems thinking methods across multiple sociotechnical system domains, and he co-developed the AcciNet method with collaborators at the Centre for Human Factors and Sociotechnical Systems.

Guy H. Walker, PhD, is a Professor within the Centre for Sustainable Road Freight at Heriot-Watt University in Edinburgh. He lectures on transportation engineering and human factors. He is the author or co-author of over 100 peer-reviewed journal articles and 18 books. He has been awarded the Institute for Ergonomics and Human Factors (IEHF) President's Medal for the practical application of ergonomics theory and Heriot-Watt's Graduate's Prize for inspirational teaching. Dr Walker has a BSc Honours degree in psychology from the University of Southampton and a PhD in human factors from Brunel University. He is a fellow of the Higher Education Academy, a fellow of the Chartered Institute of Ergonomics and Human Factors, and a member of the Royal Society of Edinburgh's Young Academy of Scotland. His research has featured in popular media from national newspapers and television and radio programmes through to an appearance on the Discovery Channel.

Paul M. Salmon, PhD, is a Professor in human factors at the Centre for Human Factors and Sociotechnical Systems at the University of the Sunshine Coast. Paul has over 20 years' experience of applied human factors and systems thinking research in areas such as road and rail safety, aviation, defence, sport and outdoor recreation, healthcare, workplace safety, and cybersecurity. His research has focused on understanding and optimising human, team, organisational, and system performance through the application of human factors and systems theory and methods. He has co-authored 22 books, over 300 peer-reviewed journal articles, and numerous book chapters and conference contributions. According to *The Australian Research Magazine*, since 2020 Paul has been Australia's number-one researcher in the field of quality and reliability. Paul's contribution has been recognised through various accolades, including the Chartered Institute for Ergonomics and Human Factor's 2019 William Floyd Award and 2008 President's Medal, the Human Factors and Ergonomics Society Australia's 2017 Cumming Memorial Medal, and the International Ergonomics Association's 2018 Research Impacting Practice Award.

PREFACE

As you read this Preface, somewhere in the world an athlete is nursing an injury, a team is grappling with performance issues, and a sports organisation is facing either operational, financial, or ethical challenges. These are not isolated incidents, they are symptomatic of a larger, more intricate problem. Sport, in all its technical skill, glory, and passion, is fundamentally a complex system (Salmon & McLean, 2021). Yet, ironically, traditional approaches to solving sports myriad issues often ignore this foundational complexity.

The Folly of Simplistic Solutions

The application of quick and simple fixes in sport is commonplace (McLean et al, 2019). When an athlete is injured, the focus is directed to rehabilitation, often without consideration of the myriad contributing factors such as athlete psychology, training programmes, and even organisational culture. When a team's performance declines, often the immediate reaction is to change the coach, or invest in new and expensive players (where possible), without attempting to understand broader systemic issues like team functioning, organisational factors, or club and executive board strategy. When sports organisations face issues, e.g. financial, ethical or legal such as doping, gambling or match fixing, the solutions are often targeted at the easiest and most obvious fixes that address problem symptoms rather than the underlying causes. The term 'fixes that fail' (Meadows, 1997) refers to the unintended consequences of solutions that are implemented without a comprehensive understanding of the systemic factors that are driving issues. These fixes that fail are not merely ineffective, they can worsen the problems they aim to solve, or create additional problems. For example, rushing an athlete to return prematurely from an injury could lead to further health issues, negatively impacting the athlete, the team, and the club long term. This cyclical pattern of problem-solving without considering the complexity associated with sport leads to recurring issues we often see in sport.

The Imperative of Systems Thinking Methods

It is not enough to merely recognise that sport is complex, it is important we equip ourselves with the tools to navigate and respond to this complexity. Traditional sports science methods, rooted in reductionism (McLean et al., 2021), are often ill-suited to understanding the web of interrelated components that constitute the sport system. What is needed is a paradigm shift towards systems thinking, a theoretical approach that embraces the interconnected, dynamic, and often non-linear nature of sport. Critically, the philosophy provides models and methods which enable researchers and practitioners to not only understand this complexity but also exploit it when developing and introducing interventions. By understanding the systemic relationships that give rise to problems, we can design interventions that are not just palliative but transformative.

We see this book as the spark that will ignite the fire for this much-needed paradigm shift. This book is an endeavour to bridge the gap between recognising the complexity inherent in sport and doing something meaningful about it. It synthesises nearly a decade of research from the authors in applying systems thinking-based methods to address problems in sport. We aim to provide not just theoretical insights but also practical guidance for athletes, coaches, administrators, sports science practitioners, and academics who are on the front lines of these complex challenges. As you read the ensuing chapters, you will encounter a blend of theory, methodology, and case study examples that demonstrate the power of systems thinking in addressing some of the most pressing and complex issues in sport today, from injuries and performance decline to organisational, ethical, and technological issues. We invite you to join us on this intellectual journey, one that challenges conventional sport research and practice and opens new avenues for understanding, intervention, and ultimately, improvement in the intricate and complex world of sport.

Welcome to the new playbook for understanding and responding to complexity in sport.

How to Read This Book

This book is intended to act as a practical guide for those wishing to apply *Systems Thinking Methods in Sport*. We expect that for most readers there will be much new information, particularly with the application of systems thinking theories and methods discussed in this book. We have tried to achieve a balance in the level of detail provided and where possible refer the novice reader to other texts that will be useful for further guidance, discussion of relevant theories, and example applications. While some methods applied in this book are related, and we encourage their integration, we have written each chapter as a standalone guide, so that readers will not be required to continually refer to other chapters. Lastly, readers will no doubt differ in specific interests; however, it is recommended that Chapters 1 and 2 be read by every reader, regardless of field or experience. These provide an overview of the theoretical platform for the methods described throughout this book. Following this, the choice of chapters should be made based on the readers' interest in specific methods, case studies, and application areas.

This book is divided into four sections:

1 Introductory chapters to complexity and systems thinking concepts (Chapters 1 and 2)
2 Systems analysis and design methods (Chapters 3–8)
3 Systemic risk and accident analysis methods (Chapters 9–12)
4 Many model applications (Chapter 13)

References

McLean, S, Read, G.J.M, Hulme, A, Dodd, K, Gorman, A.D, Solomon, C, Salmon, P.M. (2019). Beyond the tip of the iceberg: Using systems archetypes to understand common and recurring issues in sports coaching. *Frontiers in Sport and Active Living*. doi:10.3389/fspor.2019.00049

Mclean, S., Kerherve, H. Stevens, N., Salmon, P.M. (2021). A systems analysis critique of sports science research. *International Journal of Sports Physiology and Performance*.

Meadows, D. H. (2008). *Thinking in systems: A primer*. Chelsea Green Publishing, Whit River Junction, Vermont.

Salmon, P. M., & McLean, S. (2020). Complexity in the beautiful game: Implications for football research and practice. *Science and Medicine in Football*, 4(2), 162–167.

ACKNOWLEDGEMENTS

The authors would like to acknowledge and thank the numerous colleagues we have collaborated with over the years. Your pursuit of knowledge and research excellence has been a constant source of inspiration and your intellectual contributions across numerous research projects have significantly enriched the thinking that has gone into this book.

PART 1

Introductory Chapters to Complexity and Systems Thinking Concepts

1

INTRODUCTION TO COMPLEXITY AND SYSTEMS THINKING IN SPORT

> If complexity is the problem, then systems thinking is the answer.
>
> Senge (1990)

For as long as professional sport has existed it has largely been viewed, governed, and practised through a reductionist mindset, where complex challenges are broken down into individual components, thought to be easier to understand and resolve. Take, for example, the common strategy of sacking a coach following a series of poor performances. While this action might offer short-term respite, and even recovery (de Dios Tena & Forrest, 2007), it often fails to address deeper, systemic issues like organisational culture, player morale, or even medical and psychological support systems for athletes (McLean et al., 2019). Underperformance of a team is likely influenced by numerous interconnected factors beyond the coach. However, the solutions are often isolated, limited, and knee-jerk reactions. As a result, in most cases performance returns to its original state, and in some cases, it degrades even further (Flores, Forrest, & Tena, 2012). Within systems thinking, this is a well-known response that is termed 'fixes that fail' – whereby a quick but inappropriate fix is applied. While the problem symptom is temporarily alleviated, unintended consequences emerge and the problem symptom either returns or worsens (Senge, 1990). These 'fixes that fail' are ubiquitous throughout sport contexts (McLean et al., 2019; 2021).

Injury prevention represents a core area of sports science research and practice. Traditional injury management has focused on immediate treatment and rapid rehabilitation. Injured players are often segregated, subjected to isolated physical regimens, and rushed back to the field. While this may address the immediate physical dimensions of the injury, it ignores the complex network of factors that may have contributed to the injury in the first place (Hulme et al., 2017). Factors like medical history, training load, match scheduling, nutrition, psychological stress, personal issues, and even team and coaching staff dynamics may play a role. Moreover, the systemic impact of an injury on team morale, strategies, and even fan engagement is rarely considered. Instead of looking at an injury

DOI: 10.4324/9781003259473-2

as a symptom of broader issues within the complex sports system (McLean et al., 2021), the traditional model often treats it as an isolated event, missing the opportunity for broader analysis and more effective, systemic solutions.

Athlete performance has also been analysed by reductionist thinking. The prevalent mindset has often been to break down performance into discrete components and quantifiable metrics – speed, strength, endurance, passing, shots on goal, etc. (Wright et al., 2014). The focus is then on enhancing these individual components in isolation, overlooking how they interrelate and contribute to overall performance. This 'reductionist' approach makes the false assumption that complex systems can be broken into parts, the parts can then be optimised, and the system can then be put back together and will function more effectively (McLean et al., 2019). This approach fails to account for emergent properties and the complex interplay of physiology, psychology, team dynamics, and even the influence of external factors like fan support and organisational culture. An athlete's performance cannot be understood in isolation without considering their role within the team, their mental state, their relationship with coaching staff, and the strategies employed during play, among others. Overemphasising any single metric like speed or strength can create imbalances and unintended consequences elsewhere, which may lead to underperformance or even injury. The fixes derived from such an approach will likely fail.

Organisational issues in sport are also far too often approached through a reductionist lens. Budgets are cut, departments are restructured, and policies are modified in isolation without understanding how these components interact within the broader system. The result is a series of band-aid solutions that may temporarily relieve symptoms but fail to address the systemic issues which gave rise to the issues in the first instance.

When it comes to sport governance, the reductionist mindset appears yet again, often with implications that span athletes, teams, and entire sporting events. A quintessential example of this is the approach taken by anti-doping authorities. Founded with the noble goal of promoting clean sport, anti-doping authorities have focused their efforts on a three-pronged strategy: deterrence, detection, and enforcement (WADA, 2022). While this might seem like a logical and effective approach, it fundamentally fails to consider the complex systemic factors that contribute to the decision to dope in the first place (McLean et al., 2023). Factors like competition, the pressure to perform, and even the economic incentives offered for winning can push athletes towards making such choices. In some instances, national pride, as propagated by governing bodies, can add another layer of pressure that influence athletes down this precarious path. There's also the influence of coaches, medical staff, and sometimes even collusion at an organisational level that contribute to a culture where doping may be seen as necessary or even normal.

Sport is also continually evolving in ways that are increasing complexity. More and more advanced forms of technology such as artificial intelligence continue to be introduced to support athletes, coaches, and sports officials (Chmait & Westerbeek, 2021; Claudio et al., 2019; Frevel et al., 2022). Sophisticated data collection and analysis systems are also increasingly being employed in areas such as performance analysis, injury management, and training and coaching development (e.g., Goes et al., 2021; Rein & Memmert, 2016). Finally, the financial sums on offer continue to expand dramatically, which in turn enhances competitiveness and introduces ever more stakeholders seeking to turn a profit. A business-as-usual approach to understanding and optimising sports performance and athlete health and wellbeing will no longer suffice. This book represents the culmination of nearly a decade of our research applying complexity and systems thinking-based methods in sport. This is based on the notion that, as sport systems are complex in nature, systems thinking-based methods are required to describe and understand entire sport systems, their interrelated components, and emergent properties (Salmon & McLean, 2020). It is our view that, without such analyses, it is not possible to design and introduce appropriate and effective interventions. Despite this, the application of systems thinking methods in sport is lagging behind other areas. While the impetus for this book is to provide practical guidance on the application of systems thinking-based methods in sport, it is important to first orient the reader around the key underlying philosophies of complexity and systems thinking and clarify why sport is complex and necessitates a systems-thinking approach.

Sport as a complex system

Sport is a complex and dynamic system that involves a multitude of components that interact and interrelate in ways that are often difficult to predict or control (Salmon & McLean, 2020). For example, the interactions of the physiological and psychological components of athletes to the social, economic, and political influences that shape the sports industry more broadly (Salmon et al., 2021). Moreover, the interactions between these components are not linear or straightforward, and they can produce emergent properties and behaviours that are difficult to anticipate or manage. In light of this complexity, it has become increasingly important to adopt a systems thinking approach to the design and analysis of sport, one that acknowledges the interconnectedness and interdependence of the various components of the sports system.

Within scientific research there is an increasing attention on complex systems problems, yet the application of systems thinking methods is lagging (Leveson, 2017). Systems thinking methods are different from 'deterministic' methods. Deterministic methods, broadly speaking, focus on reducing systems down to their component parts on the assumption that if individual components can be understood and improved, then so can the overall system. For example, a deterministic approach to understanding sports injuries would investigate isolated components such as training load, or athlete intrinsic or extrinsic risk factors to injury such as biomechanics or physiology.

Deterministic methods have driven scientific research in many fields for decades (Dekker, 2011). They are not, however, appropriate for tackling all classes of problems. Problems such as climate change adaptation, food and water security, geopolitical instability, ecological sustainability, global economics, and public health issues are considered complex in that they are made up of a large number of heterogeneous components; these components interact with each other; the interactions produce emergent effects that are different from the effects of the individual components alone; and these effects persist over time and adapt to changing circumstances (Luke & Stamatakis, 2012). In the context of sports injuries, a systems thinking approach would attempt to understand and consider the dynamic interactions between societal influences, organisational factors, sports systems structures (e.g., policy and procedures), influences from peers, family and friends, competition influences, and individual characteristics, among others. By considering the broader system and the interactions between its components, a richer and more comprehensive understanding of both the positive and negative aspects of sport can be achieved.

Characteristics of complex systems

Complexity science is concerned with attempting to understand and respond to problems that are dynamic and unpredictable, multi-dimensional, and comprise various interrelated components. Fundamental to complexity science, and systems thinking, is a focus on the interactions among components within the 'complex system', rather than on the role and contribution of those components in isolation (Batty & Torrens, 2005; Ottino, 2003; Senge, 2006). As such, complexity is both elusive and difficult to define (Cilliers, 1998), however, various authors have outlined the characteristics exhibited by complex systems (e.g., Von Bertalanffy, 1968; Cilliers, 1998; Holland, 2014; Skyttner, 2005). For the current chapter, we have drawn upon a widely accepted set of characteristics of complexity outlined by Cilliers (1998). These are presented in Table 1.1 along with relevant examples in a sporting context.

The characteristics described in Table 1.1 can be found in all sports, regardless of level, competition, gender, age-group, or jurisdiction. This inherent complexity is at loggerheads with the ideologies of the world of sport that is focused on dissecting reality into ever-smaller parts. Therefore, to truly understand sports in all its unpredictability, we must understand the limitations of the lenses of reductionism and adopt a holistic, systems-thinking approach. This broader perspective not only offers richer insights but opens new avenues to explore in enhancing all aspects of sport. Therefore, as we delve further into the complexity of sport in this book, it is important to arm ourselves with the theories and methods that can cope with this inherent complexity of sport.

TABLE 1.1 Complex systems characteristics applied to sport

Complex system characteristic	*Definition*	*Example in sport*
Multiple interacting components	Complex systems consist of a multitude of components or agents that interact with each other. The diversity and number of these components contribute to the system's complexity, making it more than just the sum of its parts.	In a soccer team, players, coaches, tactics, support staff, and even fans interact in complex ways to influence the outcome of a match. Each player's skills contribute to the team's performance but in combination, these components may lead to unexpected victories or defeats.
Non-linear interactions and emergence	Interactions between components are abundant and can be non-linear in nature, meaning that there is an asymmetry between input and output, and small events can produce large outcomes and vice versa (Dekker, 2011). Emergent properties arising from interactions mean that 'the action of the whole is more than the sum of its parts' (Holland, 2014).	In basketball, a single three-point shot can dramatically shift the momentum of the game, and lead to emergent phenomena like a sudden boost in team morale, which might, in turn, affect the overall performance of the team positively.

(*Continued*)

TABLE 1.1 (Continued)

Complex system characteristic	*Definition*	*Example in sport*
Open system	Complex systems are open systems. This means that it is difficult to define their boundary and that they interact with, and influence their environment while being influenced by their environment in return.	Golf tournaments are influenced by a myriad of external factors such as weather conditions, terrain, and crowd behaviour. This makes them open systems, with boundaries that are difficult to define.
Ignorance of components	Components within the system are ignorant in that they respond only to local information and do not fully comprehend the behaviour of the overall system or the effects of their actions on the behaviour of the overall system.	Anti-doping authorities often do not know about unknown substances or new methods of doping that have not yet been identified or thoroughly researched. This ignorance of components adds an additional layer of complexity to the task of ensuring fair play and can have wide-ranging implications on the integrity of sports competitions.
Path dependence	Cilliers argues that complex systems have a history or path dependence. Their past is co-responsible for their present behaviour in that decisions and actions made previously (even many years previously) influence the here and now. This characteristic is referred to by others as sensitive dependence on initial conditions (Dekker, 2011).	The history of sport science has been shown to be driving multiple current behaviours within sports science. Many of the values or approaches adopted during the early years of sport science research are continuing to drive behaviour in a manner that may be detrimental to outcomes. These include the value placed on quantifiable data, reductionism, and linear thinking (McLean et al., 2021).
Information received primarily from neighbours	Cilliers (1998) describes how information received by components mainly derives from neighbouring components and how long-range interactions are limited. However, as components often interact with many other components, it is possible to influence non-neighbouring components through just a few interactions.	In rowing, team members primarily rely on immediate neighbours for pace and rhythm. This local interaction influences the overall performance of the team.

(*Continued*)

TABLE 1.1 (Continued)

Complex system characteristic	*Definition*	*Example in sport*
Recurrent loops in interactions	The effect of an activity can feed back onto itself either directly or through other components. These feedback loops can be positive or negative, and both are necessary.	Cyclists often use real-time performance metrics, e.g., speed, cadence, or power outputs to adjust their race strategy. This feedback loop can either improve or degrade performance, depending on various factors like current physical condition or environmental factors.
Dynamicity	Complex systems are dynamic, meaning they change over time. This could be due to internal factors, such as adaptation or evolution, or external factors like environmental changes. The dynamic nature of these systems makes them inherently difficult to model with static methods.	In team invasion sports, the strategies are highly dynamic, constantly adjusting to the opponent's actions and to real-time events such as injuries or penalties. As such, team invasion sports are vivid examples of the dynamic nature of complex systems.

Complexity science

There are numerous sub-fields of complexity science, e.g., dynamical systems theory, complex systems theory, cybernetics, graph theory, systems dynamics, and artificial intelligence, to name a few (see Castellani, 2018). Fundamental to complexity science is systems thinking with a focus on the interactions between the components within the system. This book does not focus on one sub-discipline of complexity, but offers a suite of theories and methods that can be used to understand and optimise individual, team, organisational, and system performance across sport. The ensuing methods are particularly suited for dissecting the intricate interplay between athletes, their equipment, and the environment, and for understanding team dynamics. Moreover, they can guide the design and evaluation of sports-related products, training programmes, and even regulatory policies, adding a layer of knowledge that will complement traditional sports research.

Case study: a systems thinking perspective on the Lance Armstrong doping scandal

This case study was adapted from The Conversation- More than one bad apple: a systems view on the Lance Armstrong doping saga (Salmon, 2015).

For this case study, we look at the Lance Armstrong doping scandal through a systems thinking lens to highlight how the overall system can contribute to system performance and individual behaviours, both good, or bad.

The Lance Armstrong doping scandal is one of the most controversial events in sports history. Armstrong, a celebrated road cyclist and seven-time Tour de France winner, was accused multiple times of using performance-enhancing drugs. After years of vehement denial, in 2012, he was stripped of his Tour de France titles and banned from professional cycling for life following a report from the US Anti-Doping Agency (USADA). This report presented substantial evidence of Armstrong's involvement in what was described as the most sophisticated, professionalised, and successful doping programme the sport had ever seen. In 2013, Armstrong confessed to doping in a televised interview with Oprah Winfrey, marking a dramatic fall from grace for a sports icon who had previously been a symbol of perseverance and triumph, especially after his recovery from testicular cancer. Following the now famous interview, Armstrong admitted he would take performance-enhancing substances again:

> *You take me back to 1995, when it was completely and totally pervasive, I'd probably do it again.*
>
> *Lance Armstrong*

As described earlier in this chapter, systems thinking reveals that behaviour is an emergent property of overall systems; that there is a shared responsibility for all behaviour; and that this responsibility is shared across all of the people within the system in which the behaviour occurs. This approach has proven powerful for identifying the systemic factors that combine to cause accidents and catastrophes. In contemporary systems thinking science, the focus is on the overall system and the network of contributory factors, rather than the decisions and actions of individuals.

A systems approach to doping

Let's now apply this thinking to Armstrong's doping transgressions. Is it acceptable to make Lance Armstrong the scapegoat? Or is it the overall elite international cycling system – comprising multiple people, organisations, and agencies – that should be held to account?

American psychologist Philip Zimbardo famously discussed 'bad apples' (or bad individuals) and 'bad barrels' (external factors that create bad behaviours) when examining wrongdoing. It is worth looking at the 'system' within which Lance Armstrong committed his doping. Was it really a case of one bad apple or were there bad barrels as well? Were there bad barrel-makers, bad barrel-owners, and bad growers too? This is not to defend Armstrong's doping, or his behaviour towards the individuals unfortunate to be involved. Rather, to improve behaviour you have to focus on the system itself, not merely the individuals within it. Focusing all of the attention on Armstrong, or any athlete, makes it difficult to prevent doping moving forward as the systemic conditions may remain.

Cycling as a complex system

The elite cycling system is complex. It is made up of multiple people, groups, agencies, and organisations, including cyclists, coaches, sports scientists, doctors, cycling teams, manufacturers, sponsors, sports organisations, governing bodies, manufacturers, and doping agencies – to name only a few. This has important implications for causality and responsibility.

First, the doping programme adopted by Armstrong and his teams was a result of interactions (good or bad) between all of these actors. Either knowingly or unknowingly, the behaviour of those within the elite cycling system enabled the doping programme – or, in some cases, supported it. Armstrong has alleged, for example, that Union Cycliste Internationale (cycling's world governing body) President Hein Verbruggen covered up a positive test in 1999. Were these interactions preventative, or if there were more of a culture of preventing the use of performance-enhancing drugs, the doping programme could not have prospered.

Second, the doping programme was enabled through multiple factors, not just Armstrong's transgressions (even if he was a driving force). A weak anti-doping system at the time is one example. Thus it is the overall system that should share the responsibility for doping, not just Armstrong himself. For example, remove Armstrong and many of the other factors that enabled him and his team to dope, and doping would likely remain. It is naive to think others would not dope within a system that could not prevent it. To this day, positive tests continue in elite cycling, indicating that the system is such that teams and cyclists will still chance their arm.

A shared responsibility for doping

So, who should share the responsibility? First, there are those who were knowingly involved: Armstrong, his teammates, doctors, coaches, and so on. The United States Anti-Doping Agency's (USADA) 'reasoned decision' on Armstrong labels the period in which Armstrong doped as the 'dirtiest ever'. The USADA report mentions peloton's code of silence – the 'omerta' – on doping. It notes that 20 of the 21 Tour de France podium finishers between 1999 and 2005 – Spanish rider Fernando Escartin is the exception – have been linked to doping. Add to this the intricate teams of doctors and suppliers needed and an intricate web of actors quickly emerges.

Second, there are those within the cycling system who were not engaged in doping programmes but were nonetheless part of the system in which it was able to continue. This includes those involved in testing, governance, coaching, and team management. They can all share some of the responsibility, even though they were not knowingly involved or doing anything wrong. This is just as aircraft designers, air traffic controllers, airlines, and others share the responsibility for aviation safety.

This all paints a picture of a broken system in which doping was able to prosper. It is hard from this viewpoint to have one scapegoat. Yes, Armstrong committed terrible acts, broke the rules, and ultimately damaged the sport's reputation. But the system itself should share some responsibility, not Armstrong alone, simply because it was able to work for so long.

The way forward

The wrong approach is to have scapegoats and to punish individuals. The correct approach is to fix the system: to identify all of the factors and their interactions across all levels of the system that in any way enabled doping to occur. These can be factors related to governing bodies, rules and regulations, coaching and training, team management, drug testing policy and procedures, culture, sponsorship, and prizemoney. Interventions should then focus on removing or strengthening these factors. Encouragingly, this is already happening through the efforts of anti-doping authorities globally. Only through in-depth, top-to-bottom systems thinking-based examinations can true systems reform happen.

Conclusion

This opening chapter aimed to set the scene for the remainder of the book. To summarise:

- We have introduced the notion that decision and actions within sport are often driven through reductionist mindset, that bring about 'Fixes that Fail'.
- We have introduced the concept of sport as a complex system that comprises multiple interacting components that produce emergent effects that are different from the effects of the individual elements alone.
- We have shown that numerous characteristics of complexity are inherent within sport, thus demonstrating that sport is indeed a complex system requiring appropriate approaches to understand it.
- We have highlighted that the structure and functioning of the system influence behaviours within that system.

References

Batty, M., & Torrens, P. M. (2005). Modelling and prediction in a complex world. *Futures*, 37, 745–766.

Bertalanffy, L. v. (1968). *General systems theory*. New York: George Braziller Inc.

Castellani, B., & Gerrits, L. (2018). *Map of the complexity sciences*.

Chmait, N., & Westerbeek, H. (2021). Artificial intelligence and machine learning in sport research: An introduction for non-data scientists. *Frontiers in Sports and Active Living*, 3, 363.

Cilliers, P. (1998). *Complexity and postmodernism*. Boca-Raton, FL: Routledge.

Claudino, J. G., Gabbett, T. J., de Sá Souza, H., Simim, M., Fowler, P., de Alcantara Borba, D.,... & Nassis, G. P. (2019). Which parameters to use for sleep quality monitoring in team sport athletes? A systematic review and meta-analysis. *BMJ Open Sport & Exercise Medicine*, 5(1), bmjsem-2018.

de Dios Tena, J., & Forrest, D. (2007). Within-season dismissal of football coaches: Statistical analysis of causes and consequences. *European Journal of Operational Research*, 181(1), 362–373.

Dekker, S. (2011). *Drift into failure: From hunting broken parts to understanding complex systems*. Ashgate, Vermont.

Flores, R., Forrest, D., & Tena, J. D. (2012). Decision taking under pressure: Evidence on football manager dismissals in Argentina and their consequences. *European Journal of Operational Research*, 222(3), 653–662.

Frevel, N., Beiderbeck, D., & Schmidt, S. L. (2022). The impact of technology on sports–A prospective study. *Technological Forecasting and Social Change*, 182, 121838.

Goes, F. R., Meerhoff, L. A., Bueno, M. J. O., Rodrigues, D. M., Moura, F. A., Brink, M. S.,... & Lemmink, K. A. P. M. (2021). Unlocking the potential of big data to support tactical performance analysis in professional soccer: A systematic review. *European Journal of Sport Science*, 21(4), 481–496.

Holland, J. H. (2014). *Complexity: A very short introduction*. Oxford: Oxford University Press.

Hulme, A., Salmon, P., Nielsen, R. O., Read, G. J., & Finch, C. (2017). From control to causation: Validating a 'complex systems model' of running-related injury development and prevention. *Applied Ergonomics*, 65, 345–354.

Leveson, N. G. (2017). Rasmussen's legacy: A paradigm change in engineering for safety. *Applied Ergonomics*, 59, 581–591.

Luke, D. A., & Stamatakis, K. A. (2012). Systems science methods in public health: Dynamics, networks, and agents. *Annual Review of Public Health*, 33, 357.

McLean, S., Naughton, M., Kerhervé, H., & Salmon, P. M. (2023). From Anti-doping-I to Anti-doping-II: Toward a paradigm shift for doping prevention in sport. *International Journal of Drug Policy*, 115, 104019.

McLean, S., Kerhervé, H. A., Stevens, N., & Salmon, P. M. (2021). A systems analysis critique of sport-science research. *International Journal of Sports Physiology and Performance*, 16(10), 1385–1392.

McLean, S., Coventon, L., Finch, C. F., & Salmon, P. M. (2022). Incident reporting in the outdoors: A systems-based analysis of injury, illness, and psychosocial incidents in led outdoor activities in Australia. *Ergonomics*, 65(10), 1421–1433.

McLean, S., Read, G. J., Hulme, A., Dodd, K., Gorman, A. D., Solomon, C., & Salmon, P. M. (2019). Beyond the tip of the iceberg: using systems archetypes to understand common and recurring issues in sports coaching. *Frontiers in Sports and Active Living*, 1, 49.

Ottino, J. M. (2003). Complex systems. American Institute of Chemical Engineers. *AIChE Journal*, 49(2), 292.

Peter, S. (1990). *The fifth discipline*. New York: The Art & Practice of Learning Organization. Doubleday Currence.

Rein, R., & Memmert, D. (2016). Big data and tactical analysis in elite soccer: Future challenges and opportunities for sports science. *SpringerPlus*, 5(1), 1–13.

Salmon, P. M., & McLean, S. (2020). Complexity in the beautiful game: Implications for football research and practice. *Science and Medicine in Football*, 4(2), 162–167.

Skyttner, L. (2005). *General systems theory: Problems, perspectives, practice*. Singapore: World Scientific.

World Anti-Doping Agency. (2022). The world anti-doping code. Retrieved from https://www.wada-ama.org/en/what-we-do/world-anti-doping-code

Wright, C., Carling, C., & Collins, D. (2014). The wider context of performance analysis and it application in the football coaching process. *International Journal of Performance Analysis in Sport*, 14(3), 709–733.

2
WHERE TO INTERVENE IN SPORTS SYSTEMS

Background

The broad aim of sports research is to understand and enhance the performance, health, and/or wellbeing of coaches, athletes, clubs, leagues, organisations (McLean, Kerhervé et al., 2021; McLean, Rath et al., 2021). With the increased participation and interest in organised sport throughout the 21st century has come a proliferation of academic research attempting to understand every aspect of sport and its broader societal influence. This has led to a myriad of changes and improvements in sport. For instance, the male marathon world record has witnessed a remarkable evolution from 2:32:35 in 1920 to 2:00:35 in October 2023 as the cumulative result of improvements in training, nutrition, shoe technology, running surface technology, and athlete psychology (among other factors) (Goss et al., 2022; Hoogkamer et al., 2017). An example of an organisational improvement is the formulation of the World Anti-Doping Agency (WADA), a hybrid organistion of national governments and the Olympic movement which was created with a core mission to harmonise anti-doping policy globally (Read et al., 2020).

As demonstrated in Chapter 1, sport in all its varied contexts are complex systems. This inherent complexity can be difficult to define. Nonetheless, complex systems typically contain structurally and functionally different components which interact in a non-linear, dynamic manner to produce emergent behaviours (Cilliers, 2002; Salmon & McLean, 2020). Complex systems exist within multiple levels of sport, depending on where the boundary conditions are drawn, and these can include individuals, matches, teams, sporting leagues, major events (e.g., the Olympics), and other entities such as governance (e.g., WADA, National Sporting Organisations [NSOs]). In any complex system, there are places where we can intervene in an attempt to change behaviour and, ideally, system outcomes.

The complexity inherent within sport renders decisions about where to intervene to make fundamental improvements, difficult. The current thinking in sport often seeks interventions which are aimed at fixing or solving singular broken components within

DOI: 10.4324/9781003259473-3

the system (e.g., the athlete, the equipment, and the resources). Meadows (1997) seminal work in systems thinking neatly describes 12 different points of where to intervene in complex systems (termed 'Leverage Points'). Leverage Points are ordered into a hierarchy based on their potential effectiveness to influence transformational change to a system. Deeper Leverage Points (e.g., the power to change mindsets/mental models and the structure of the system) have a stronger influence on system behaviour and performance, while shallow Leverage Points (e.g., system numbers, parameters, constants) have less impact (Meadows, 1997). For example, a shallow Leverage Point, such as adjusting the number of hours an athlete spends on different aspects of training (strength vs. skill training, for example), can significantly impact performance, but will not result in transformational systemic change to the sport. Whereas, shifting from viewing athletes as mere performers to seeing them as holistic individuals who require physical and mental health support represents a transcendent change in sports paradigms, that would have transformational systemic change.

Implicit within the Leverage Points hierarchy is the understanding that influencing the deeper points subsequently alters each of the shallower points. Building off of Leverage Points, Abson et al. (2017) suggested these points could be overlayed with four systems 'Realms of Leverage' (arranged from deeper to shallow): (1) mental models, (2) system design/structure, (3) feedback loops, and (4) parameters, as described in Table 2.1.

Leverage is counterintuitive

One of the most interesting principles of systems thinking described by Meadows (1997) is the counterintuitive nature of true leverage points. Often, what appears to be an obvious solution to a problem may only address symptoms, rather than underlying causes. The real Leverage Points, the places where a small shift can produce a disproportionate impact are often not intuitive and can often be overlooked. Take injuries in football, arguably influenced by congested fixture schedules due to financial incentives of TV broadcast deals. The immediate, and somewhat intuitive response might be to invest more in medical staff or state-of-the-art recovery technologies and methods for athletes. While this could yield some benefits, it is essentially a reactive measure that doesn't address the underlying cause, which is the overburdening schedule itself. The counterintuitive Leverage Point here could be in altering the rule structure around scheduling, as mentioned earlier. It doesn't necessarily give the immediate financial gains or even the entertainment value that TV broadcast deals promise, but it offers something far more sustainable in a healthier, more competitive league, with increased athlete availability, and athletes performing at their peak. Therefore, the Leverage Point lies not in merely treating or managing injuries better but in changing the systemic conditions that lead to those injuries in the first place. The concept of counterintuitive Leverage Points underscores the importance of deeply understanding the system's structure and behaviour (Meadows, 1997). In the case of reducing injuries in professional sport, what might initially appear to be sacrificing profit or entertainment, could in the long run enhance the sport's overall value, quality, and even its financial capability. It may even result in improved profit or entertainment long term. While the example here is simplified, it often takes a systems-thinking lens to identify such Leverage Points, revealing that the most effective solutions often lie where we least expect them.

TABLE 2.1 The Realms of Leverage (Abson et al., 2017), Leverage Points (Meadows, 1997), and specific examples of the different Leverage Points from sport

Realms of Leverage (Abson et al., 2017)	*Leverage Points (LP) (Meadows, 1997)*	*Examples from sport*
Parameters The mechanistic, immediate characteristics of the system.	Constants, parameters, numbers	Adjusting the number of hours an athlete spends on different aspects of training (strength vs. skill training, for example) can significantly impact performance.
	Buffer sizes	The depth of a team's roster serves as a buffer, allowing for better performance even when key players are injured.
	Material stocks and flows	The quality and availability of training facilities can drastically affect an athlete or club's performance over a season.
Feedback loops The interactions between elements within a system of interest.	Relative delays	Investment in youth development may take years to show results in the form of talent progressing to the elite level.
	Negative feedback loops	Anti-doping regulations and testing create a feedback loop intended to balance and maintain fair competition.
	Positive feedback loops	A successful athlete gains more media exposure, leading to more sponsorship opportunities, which in turn can be invested back into training – creating a reinforcing loop for success.
System design/structure The structures and rules in place that manage the feedback loops and parameters.	Information flows	Proprietary analytics and scouting information can give teams a significant advantage over their competition.
	Rules of the system	Salary caps or luxury taxes can drastically change team-building strategies.
	Self-organising system structures	An athlete's ability to adapt and evolve their training regimen, techniques, or specialties can shape their career trajectory.
Mental models The underpinning values, goals and world views of actors within the system which shape the emergent direction to which the system is orientated.	Goal of the system	A sports organisation with the goal of developing local talent may produce different outcomes compared to one focused solely on immediate success through purchasing expensive players.
	Mindset/paradigm	Cultural views on sports, like the win-at-all-costs mentality, can impact everything from athlete development to fan engagement.
	The power to transcend paradigms	The shift from viewing athletes as mere performers to seeing them as holistic individuals who require mental health support represents a transcendent change in sports paradigms.

Case study application: leverage points in sport research

In sport, interventions can come in different forms, from the application of new technologies, applying existing technologies to new problem spaces, new athlete training interventions, rule changes, changes in funding models, and changes in laws (Branch, 2003; Cermak et al., 2012; Gordon, 2009; Lockie et al., 2012; Meir et al., 2001; Stewart, 2017). Looking through a Leverage Points lens, it is currently unknown where the interventions used in sport research are focused, whether it be on fixing isolated components within a system, or with a whole of system focus. The aim of this case study was therefore to examine interventions in the sport research literature through the Leverage Points and Realms of Leverage frameworks. Specifically, the aim was to determine what leverage points and Realms of Leverage have been targeted by previous interventions described in highly cited sports science, sports nutrition, sports medicine, sport management, and sports law/policy literature.

Procedure

Journals were selected to represent sports science (*Journal of Sports Sciences* [JSS]), sports nutrition (*International Journal of Sport Nutrition and Exercise Metabolism* [IJSNEM]), sports medicine (*British Journal of Sports Medicine* [BJSM]), sport management (*Journal of Sport Management* [JSM]), and sports law/policy (*International Journal of Sport Policy and Politics* [IJSPP]). These journals are placed in the top quartile of their respective fields. To assess the focus of interventions in these journals, an assessment of their location on the leverage points and Realms of Leverage hierarchy (Table 2.1), was undertaken.

The top 10 articles based on their all-time citation count were downloaded for each journal listed above. Citation counts may not be indicative of how citation trends have changed over time nor are they an ideal proxy for research quality or impact on a field, but they do provide some evidence as to how often an article is being used (and therefore cited) by the wider community. As the point of intervention was of interest, these articles included only original research, with reviews (including systematic, narrative, and/or scoping reviews) excluded. Each article was then assessed by the authors by to determine which Leverage Point and accompanying Realm of Leverage which was targeted by the intervention described.

Results

Table 2.2 provides examples from the included studies at the different Leverage Points and accompanying Realms of Leverage.

The frequency of articles coded to each of the leverage points is presented in Table 2.3. Overall, across the articles assessed, 48% of the interventions were in the parameters level, while 24%, 8%, and 20% were in the feedback loops, system design/structure, and mental models levels, respectively when assessing relative to the Realms of Leverage.

TABLE 2.2 Examples of different interventions from the included studies across the Realms of Leverage (Abson et al., 2017) and Leverage Points (Meadows, 1997) frameworks

Realms of Leverage (Abson et al., 2017)	*Leverage points (Meadows, 1997)*	*Example (Reference)*
Parameters	Constants, parameters, numbers	Supplement use patterns in young German athletes (Braun et al., 2009)
	Buffer sizes	–
	Material stocks and flows	The (mis)application of economic impact analyses of sports facilities and events (Crompton, 1995)
Feedback loops	Relative delays	–
	Negative feedback loops	Methods of weight loss to make weight during weight across a range of weight categorised combat sports (Brito et al., 2012)
	Positive feedback loops	Government investment in sport and the 'virtuous' cycle wherein investment can lead to positive societal outcomes (Grix & Carmichael, 2012)
System design/structure	Information flows	Measuring the concept of 'legacy' with respect to sporting mega events (e.g., the Olympics) (Preuss, 2019)
	Rules of the system	Examining how the 'legacy' of mega sporting events influences subsequent national policy (Grix et al., 2017)
	Self-organising system structures	Inclusive youth sport structure using the 3P model (participation, performance, and personal development) (Côté & Hancock, 2016)
Mental models	Goal of the system	Why organisations support community sport and the drivers of support (Miragaia et al., 2017)
	Mindset/paradigm	The underpinning values held for brand associations in sport in three characteristics: attribute, benefit, and attitude (Gladden & Funk, 2002)
	The power to transcend paradigms	–

TABLE 2.3 The frequency of interventions which were coded to each of the Leverage Points (LP) (Meadows, 1997) and Realms of Leverage (Abson et al., 2017) frameworks from the ten most cited articles in each of the journals investigated. The largest frequency of coded articles for each of the different journals is shaded in grey

	Parameters			*Feedback loops*			*System design/structure*			*Mental models*		
	LP 1	*LP 2*	*LP 3*	*LP 4*	*LP 5*	*LP 6*	*LP 7*	*LP 8*	*LP 9*	*LP 10*	*LP 11*	*LP 12*
British Journal of Sports Medicine (BJSM)	9	–	–	–	–	1	–	–	–	–	–	–
International Journal of Sport Nutrition and Exercise Metabolism (IJSNEM)	6	–	–	–	1	3	–	–	–	–	–	–
International Journal of Sport Policy and Politics (IJSPP)	–	–	–	–	–	3	1	2	1	3	–	–
Journal of Sport Management (JSM)	–	–	1	–	–	2	–	–	–	–	7	–
Journal of Sports Sciences (JSS)	8	–	–	–	–	2	–	–	–	–	–	–

Discussion

> *Give me a lever long enough… and I shall move the world.*
>
> – *Archimedes*

The findings indicate that highly cited research in journals SM, JSS, and IJSNEM predominantly include articles with interventions which are focused at the shallow end of the Leverage Points/Realms of Leverage hierarchy (i.e., parameters, and feedback loops). This therefore indicates that the majority of research in sports science, sports medicine, and sport and exercise nutrition are focused on the Leverage Points which are likely to be the least effective in influencing broad system behaviour. In these fields there is clearly potential to design and deploy interventions which focus on deeper Leverage Points where the aim is to facilitate systemic change that will flow down to enact change at the lower levels. Let us take, for example, a football club who changes their emphasis from a strictly performance mindset which prioritises winning every match often at the cost of losing players to injury who play when not fully fit. An intervention here to shift that mindset (a deeper Leverage Point) to one which focuses on holistic player health and wellbeing would result in changes to the playing and coaching style, the rehabilitation programmes within the club, choice of players to develop and recruit, nutrition products used, and other player welfare wraparound services the club may provide.

In this case study analysis, the articles which were published in JSM and IJSPP were predominantly focused on interventions at deeper Leverage Points (i.e., system design/structure and mental models). Examples of these included the 'legacy' of major sporting events (Grix et al., 2017), the policies and practice of athletes in the university sector (Aquilina & Henry, 2010), and relationships between brand image and fan loyalty (Bauer et al., 2008). Conceptually, the focus of interventions in these domains appears inherently at influencing the system structure and collective mental models (i.e., the deeper levels of the system). The study by Grix et al. (2017) may, for example, be influencing the choices of future governments to bid for major sporting events, which in turn effect decisions on infrastructure development, jobs and careers, and decisions by professional sporting teams to use the subsequent infrastructure. As noted by Meadows (1997), to influence effective and lasting change, it is necessary to ensure that interventions are aimed at the deeper Leverage Points of the system. Therefore, when designing interventions with the aim of stimulating large-scale change, it is pertinent to include experts who have knowledge across various fields of research (i.e., relevant subject matter experts), which these articles appear to be doing. Further, inclusion of systems thinking experts, and collaboration between these experts and subject matter experts in the various research fields in sport is necessary if research translation is to lead to implementation 'at the coal face'. This collaboration and the use of systems thinking should assist in reducing the research to practise translation gap (McLean et al., 2021a).

Interventions analysed from BJSM, JSS, and IJSNEM articles were primarily focused at the shallower end of the leverage hierarchy (Table 2.1). This may be indicative of a lack of necessity in needing to design deeper interventions at the higher Leverage Points in these fields of research. Other factors which may be influencing this is the historical biases in the values, theories, and approaches that have been adopted in these fields (McLean et al., 2021a). Alternatively, it could suggest that researchers who typically publish in these journals are collecting data locally with the sporting team/environment or working directly with athletes/players who naturally become the unit of analysis. Indeed, often individuals will fill roles as both servicing practitioner (e.g., sports scientist, sports dietician, high-performance manager) and academic (i.e., the 'pracademic') to facilitate research (Collins & Collins, 2019). In this way pracademics use research to answer questions which are perceived to have direct importance to the athlete(s), coach(s), and/or team, they are working with. Conceptually, this requires that they 'work fast' to solve questions which have a direct and more immediate application (Coutts, 2016). Alternatively, interventions which focus at deeper leverage points may require a 'working slow' approach which is more thoughtful, considered, and indirect (Coutts, 2016).

This then leads to the need to consider or imagine what interventions at deeper Leverage Points might encompass. An example which focuses on shifting the goals of the system and the paradigm out of which the system arises can be found in anti-doping. Current strategies for anti-doping are primarily focused on the individual engaging in doping under the 'strict liability' legal standard, and on engineering of new tests (i.e., detection), anti-doping education of athletes and coaches (i.e., deterrence), and stricter sanctions for those detected (i.e., punishment) (Dimeo & Møller, 2018). Importantly, such interventions are at the shallow end of the Leverage Point hierarchy. In anti-doping, previous research has suggested a move from the current detection-deterrence-punishment paradigm which resembles the broader societal discourse around the 'war on drugs' to one which focuses on athlete health and harm reduction is warranted (Kayser & Broers, 2012). Such a paradigmatic change would need to be accepted by the raft of stakeholders (including athletes, coaches, administrators, and organisations), and may not be permissible to those in positions of accountability and influence within sport. This likely resistance to change makes such Leverage Points difficult to access.

This case study analysis introduces the concepts of Leverage Points and Realms of Leverage to examine research interventions across different fields of research in sport. The results of the analysis of the most highly cited research in sport indicate that journals which focus on sports medicine, sports science, and sports nutrition/exercise metabolism predominantly rely on interventions which are at the shallow end of the leverage points/ realms of leverage hierarchy (e.g., parameters, feedback loops). Conversely, journals which focus on sports management and sports law/policy were associated with interventions at deeper leverage points (e.g., system design/structure, mental models).

Overall, to achieve transformational change and maximal impact across sport, interventions might be better placed at the deeper Leverage Points. The collaborative efforts of multi-disciplinary teams in sports can lead to a paradigm shift in how sports are managed and perceived through systems thinking. By harnessing the power of interventions at deeper and shallow Leverage Points, and the expertise of diverse stakeholders, sport can evolve to meet the challenges of the 21st century, fostering environments where athletes thrive and organisations flourish.

Recommended Reading

Meadows, D. (1999). Leverage points. Places to Intervene in a System, 19, 28.

Abson, D. J., Fischer, J., Leventon, J., Newig, J., Schomerus, T., Vilsmaier, U.,... & Lang, D. J. (2017). Leverage points for sustainability transformation. Ambio, 46, 30–39.

References

Abson, D. J., Fischer, J., Leventon, J., Newig, J., Schomerus, T., Vilsmaier, U., Von Wehrden, H., Abernethy, P., Ives, C. D., & Jager, N. W. (2017). Leverage points for sustainability transformation. *Ambio*, 46(1), 30–39.

Aquilina, D., & Henry, I. (2010). Elite athletes and university education in Europe: A review of policy and practice in higher education in the European Union Member States. *International Journal of Sport Policy and Politics*, 2(1), 25–47.

Bauer, H. H., Stokburger-Sauer, N. E., & Exler, S. (2008). Brand image and fan loyalty in professional team sport: A refined model and empirical assessment. *Journal of Sport Management*, 22(2), 205 226.

Branch, J. D. (2003). Effect of creatine supplementation on body composition and performance: A meta-analysis. *International Journal of Sport Nutrition and Exercise Metabolism*, 13(2), 198–226.

Braun, H., Koehler, K., Geyer, H., Kleinert, J., Mester, J., & Schanzer, W. (2009). Dietary supplement use among elite young German athletes. *International Journal of Sport Nutrition and Exercise Metabolism*, 19(1), 97–109.

Brito, C. J., Roas, A., Brito, I. S. S., Marins, J. C. B., Cordova, C., & Franchini, E. (2012). Methods of body-mass reduction by combat sport athletes. *International Journal of Sport Nutrition and Exercise Metabolism*, 22(2), 89–97.

Cermak, N. M., Gibala, M. J., & van Loon, L. J. C. (2012). Nitrate supplementation's improvement of 10-km time-trial performance in trained cyclists. *International Journal of Sport Nutrition and Exercise Metabolism*, 22(1), 64–71.

Cilliers, P. (2002). *Complexity and postmodernism: Understanding complex systems*. Routledge.

Close, G. L., Kasper, A. M., & Morton, J. P. (2019). From paper to podium: Quantifying the translational potential of performance nutrition research. *Sports Medicine*, 49(1), 25–37. https://www.taylorfrancis.com/books/mono/10.4324/9780203012253/complexity-postmodernism-paul-cilliers

Collins, L., & Collins, D. (2019). The role of 'pracademics' in education and development of adventure sport professionals. *Journal of Adventure Education and Outdoor Learning*, 19(1), 1–11.

Côté, J., & Hancock, D. J. (2016). Evidence-based policies for youth sport programmes. *International Journal of Sport Policy and Politics*, 8(1), 51–65.

Coutts, A. J. (2016). Working fast and working slow: The benefits of embedding research in high performance sport. *International Journal of Sports Physiology and Performance*, 11(1), 1–2.

Crompton, J. L. (1995). Economic-impact analysis of sports facilities and events – 11 sources of misapplication. *Journal of Sport Management*, 9(1), 14–35.

Dimeo, P., & Møller, V. (2018). The anti-doping crisis in sport: Causes, consequences, solutions. *International Journal of Sport Communication*, 12, 434–437. Routledge.

Gladden, J. M., & Funk, D. C. (2002). Developing an understanding of brand associations in team sport: Empirical evidence from consumers of professional sport. *Journal of Sport Management*, 16(1), 54–81.

Gordon, G. (2009). Sports betting: Law and policy. A UK perspective. *The International Sports Law Journal*, (3–4), 127–132.

Goss, C. S., Greenshields, J. T., Noble, T. J., & Chapman, R. F. (2022). A narrative analysis of the progression in the top 100 marathon, half-marathon, and 10-km road race times from 2001 to 2019. *Medicine & Science in Sports & Exercise*, 54(2), 345–352.

Grix, J., Brannagan, P. M., Wood, H., & Wynne, C. (2017). State strategies for leveraging sports mega-events: Unpacking the concept of 'legacy'. *International Journal of Sport Policy and Politics*, 9(2), 203–218.

Grix, J., & Carmichael, F. (2012). Why do governments invest in elite sport? A polemic. *International Journal of Sport Policy and Politics*, 4(1), 73–90.

Hoogkamer, W., Kram, R., & Arellano, C. J. (2017). How biomechanical improvements in running economy could break the 2-hour marathon barrier. *Sports Medicine*, 47(9), 1739–1750.

Kayser, B. (2018). What might a partially relaxed anti-doping regime in professional cycling look like? In *Doping in Cycling*, 164–174. Routledge.

Kayser, B., & Broers, B. (2012). The Olympics and harm reduction? *Harm Reduction Journal*, 9, 1–9.

Kayser, B., & Smith, A. C. (2008). Globalisation of anti-doping: The reverse side of the medal. *British Medical Journal*, 337.

Lockie, R. G., Murphy, A. J., Schultz, A. B., Knight, T. J., & de Jonge, X. A. J. (2012). The effects of different speed training protocols on sprint acceleration kinematics and muscle strength and power in field sport athletes. *The Journal of Strength & Conditioning Research*, 26(6), 1539–1550.

McLean, S., Kerhervé, H. A., Stevens, N., & Salmon, P. M. (2021). A systems analysis critique of sport-science research. *International Journal of Sports Physiology and Performance*, 16(10), 1385–1392.

McLean, S., Rath, D., Lethlean, S., Hornsby, M., Gallagher, J., Anderson, D., & Salmon, P. M. (2021). With crisis comes opportunity: Redesigning performance departments of elite sports clubs for life after a global pandemic. *Frontiers in Psychology*, 11, 588959.

McLean, S., Read, G. J., Hulme, A., Dodd, K., Gorman, A. D., Solomon, C., & Salmon, P. M. (2019). Beyond the tip of the iceberg: Using systems archetypes to understand common and recurring issues in sports coaching. *Frontiers in Sports and Active Living*, 1, 49.

Meadows, D. H. (1997). Places to intervene in a system (in increasing order of effectiveness). *Whole Earth*, 1, 78.

Meir, R., Colla, P., & Milligan, C. (2001). Impact of the 10-meter rule change on professional rugby league: Implications for training. *Strength & Conditioning Journal*, 23(6), 42–46.

Miragaia, D. A., Ferreira, J., & Ratten, V. (2017). Corporate social responsibility and social entrepreneurship: Drivers of sports sponsorship policy. *International Journal of Sport Policy and Politics*, 9(4), 613–623.

Preuss, H. (2019). Event legacy framework and measurement. *International Journal of Sport Policy and Politics*, 11(1), 103–118.

Read, D., Skinner, J., Lock, D., & Houlihan, B. (2020). Balancing mission creep, means, effectiveness and legitimacy at the world anti-doping agency. *Performance Enhancement & Health*, 8(2–3), 100175.

Salmon, P. M., & McLean, S. (2020). Complexity in the beautiful game: Implications for football research and practice. *Science and Medicine in Football*, 4(2), 162–167.

Stewart, B. (2017). *Sport funding and finance*. Routledge. https://www.taylorfrancis.com/books/mono/10.4324/9780203794975/sport-funding-finance-bob-stewart

PART 2

Systems Analysis and Design Methods

3

HIERARCHICAL TASK ANALYSIS (HTA)

Background

A unified acknowledgement of sport enthusiasts, athletes, coaches, and analysts is that the world of sports is anything but simplistic. From a golfer calculating various elements like wind speed and ground slope to execute the perfect swing, to a soccer player navigating the myriad variables of taking a penalty, or a basketball team coordinating complex plays in a fast-break situation. Each sporting action is a culmination of a series of interconnected goals, sub-goals, and actions. This complexity does not stop at the moment-to-moment actions on the field or court, it extends to longer-term decisions and actions, such as training design, match tactics, periodisation, among others. In this chapter, we aim to decode and dissect task complexity to understand how it can translate into actionable insights through Hierarchical Task Analysis (HTA).

Hierarchical Task Analysis (Annett et al., 1971) is one of numerous task analysis methods, yet is arguably the most applied. Unlike the name suggests HTA does not focus exclusively on tasks, rather it focuses on goals (an objective or end state) and decomposes them hierarchically to identify sub-goals and requisite operations (Stanton, 2006). The task focused name is also misleading in terms of what can be analysed with HTA, as it can be applied at scale wherever the system boundary is drawn. This means that it can be used to analyse in-depth a sporting task, sports club operations, or even the structure of an entire sports system or major events.

The origins of HTA are in the scientific management movement of the early 20th century (Stanton, 2006). Scientific management methods were used to describe and analyse tasks in a way that supported the development of more efficient work processes. The focus was on how the work was performed physically, what was required to do the work, and how the work could be enhanced (Stanton, 2006). While this approach was successfully applied until the mid-20th century, HTA was developed in the 1960s in response to the changing nature of work. Work tasks were becoming more cognitive in nature which in turn created the need for methods which could describe both the physical and cognitive aspects of work (Annett, 2004). When introduced, HTA provided a significant

DOI: 10.4324/9781003259473-5

advancement over scientific management methods which largely focused on the physical and observable aspects of behaviour.

HTA is used to decompose systems, tasks, or behaviours into a hierarchy of goals, sub-ordinate goals, operations, and plans. HTA is used to describe and understand "what an operator is required to do, in terms of actions and/or cognitive processes to achieve a system goal" (Kirwan & Ainsworth 1992, p. 1). HTA can also be used to describe what a system is required to do in terms of the operations undertaken by multiple stakeholders to achieve systems goals. It is important to note here that an 'operator' may be human or non-human (e.g., system artefacts such as equipment, devices, documentation, and interfaces) (Salmon et al., 2022).

HTA outputs specify the overall goal of a particular task/scenario/system, the sub-goals required to achieve this goal, the operations that need to be undertaken to achieve each of the sub-goals, and the plans that dictate the order and context in which goals and operations are undertaken. The plans are key in that they can specify the interrelations between goals, which is consistent with a systems lens (Salmon et al., 2022).

The flexibility of HTA is one of the reasons for its popularity. HTA has been applied in numerous contexts for a variety of purposes (see Stanton, 2006; Salmon et al, 2022). It can be used in any domain and is often the starting point for additional forms of analysis (Salmon et al., 2010; 2022; Stanton et al., 2013). For example, in Chapter 10, the Networked Hazard Analysis and Risk Management System (Net-HARMS) (Dallat et al., 2017) risk assessment method requires a HTA as its primary data input. As a result, HTA has been applied for all manner of purposes, including interface design and evaluation, job design, training programme design and evaluation, error prediction and risk assessment, team task analysis, situation awareness requirements analysis, allocation of functions analysis, workload assessment and procedure design (Stanton, 2006; Stanton et al., 2013). While it is surprising that HTA has received less attention in sports science, there is no doubting its potential utility or breadth of possible applications.

Applications in sport

Though there are no published applications of HTA in sport, it is a generic method and can be applied in sport for many different purposes. These include describing and analysing athlete behaviour, sports club operations, and overall sports systems, and also aspects such as training design, sports product equipment design, and error identification and analysis. Outside of sport, HTA has been applied extensively in many domains ranging from defence (Stanton et al., 2010), rail (Naweed et al., 2018), healthcare (Demirel et al., 2016); Yu et al., 2014), road transport (Bedinger et al., 2015) and maritime (Ramos et al., 2019) to crime scene investigation (Smith et al., 2008), air accident investigation (Nixon & Braithwaite, 2018), and vineyard cultivation (Fargnoli et al., 2019).

Procedure and advice

A flowchart depicting the HTA procedure is presented in Figure 3.1 (adapted from Salmon et al, 2022). The HTA process is simplistic and involves collecting data regarding the task or system under analysis (through techniques such as observation, questionnaires, interviews with SMEs, walkthroughs, user trials, and documentation review

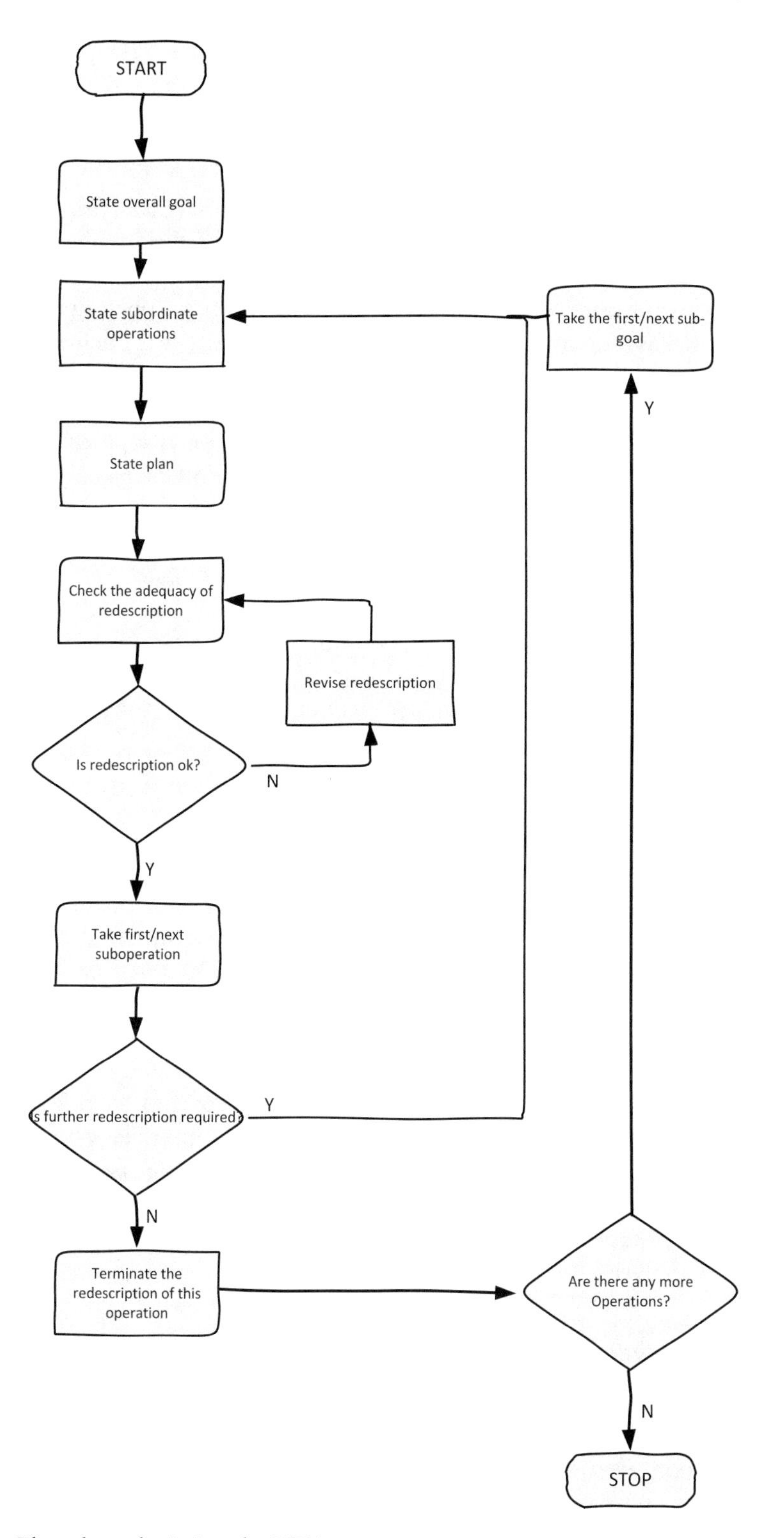

FIGURE 3.1 Flowchart depicting the HTA process.

(Salmon et al., 2022). This data is then used to identify relevant goals and sub-goals and decompose them into operations.

Step 1: Define aims of the analysis and the task or system under analysis

The first step involves clearly defining the tasks or system under analysis along with any analysis boundaries. For example, the analysis may be focused on individual behaviour in a given context (e.g., nutritional supplement use by athletes), teamwork (e.g., develop and implement match tactics), organisational behaviour (e.g., operation of a club or league), or the behaviour of an entire sociotechnical system (e.g., delivery of a major sporting event).

As part of this first step, the aims of the analysis should also be clearly defined. For example, for a HTA developed for an athlete's use of nutritional supplements (described in the case study example in this chapter) the following aims and analysis boundaries were determined during Step 1:

- Task/System under analysis: Optimise health and performance through nutritional supplement use;
- Aim of the analysis: Describe the goals, sub-goals, and operations required to deliver safe and compliant supplement use by athletes;
- Analysis boundaries: Scope of the analysis is limited to elite athletes seeking to commence the use of nutritional and dietary supplements.

Step 2: Data collection

Step 2 involves collecting relevant data regarding the task or system under analysis, with the specific data to be collected driven by the analysis aims and boundary. At a minimum, data should be collected regarding the goals and tasks involved, the human and non-human agents involved and their interactions during tasks, the ordering of tasks and contextual conditions that dictate task sequences, and information on the factors that influence behaviour. A number of different approaches can be used to collect this data with the choice often influenced by resources available, level of access (e.g., to athletes, clubs, scenarios), and project and time constraints. Useful approaches include observations, concurrent verbal protocols, structured, or semi-structured interviews (e.g., the Critical Decision Method), questionnaires and surveys, walkthrough analysis, and documentation review (e.g., incident reports, standard operating procedures) (Salmon et al., 2022). A combination of observations, interviews, and document review is typically sufficient to produce a valid analysis.

Step 3: Data transcription and analysis

Data collected through observations, interviews, and concurrent verbal protocols should be transcribed as appropriate. Analysis of the data should then proceed. This can take various forms depending on what data collection approaches are used. For example, when observations or interviews are used, the data analysis process typically involves coding the data to identify goals, sub-goals, tasks, operations, plans, and performance influencing factors.

Step 4: Determine overall goal

The analysis process begins by specifying an overall goal for the task or system under analysis. This should represent the overarching goal of the activity or system under analysis and is placed at the top of the hierarchy and labelled 0. For example, in the nutritional supplement use by athletes case study example, the overall goal was expressed as:

0 Optimise health and performance through supplement use.

Step 5: Identify and record sub-goals

The next step involves decomposing the overall goal into a series of meaningful sub-goals (according to Salmon et al. (2022) four or five is normally optimal; however, this is dependent on the task/system under analysis and can extend beyond 10). The sub-goals should be numbered, e.g., 1, 2, 3, and 4 etc. Where different sub-goals are undertaken by different actors it is useful to make a note of this within the sub-goal description.

For the nutritional supplement use by athletes case study, the overall goal of 'Optimise health and performance through supplement use' was decomposed into the following sub-goals:

1 Identify need
2 Research supplements
3 Obtain TUE
4 Acquire supplements
5 Administer supplements
6 Assess supplement efficacy

Step 6: Sub-goal decomposition

Next, the sub-goals identified during Step 5 should be broken down into further sub-goals and operations, according to how the sub-goals are achieved. It is important here to include the range of options available to achieve the different sub-goals, as opposed to describing only one ideal or typical way of achieving them. For example, in the case study example the sub-goal 'Research supplements' can be achieved by checking the banned list, checking supplements, and/or checking the batch testing register, with some athletes undertaking only one of the three operations and others undertaking all three. It is also important here to note that the HTA should reflect activity as it is actually undertaken rather than activity as it is imagined or desired (e.g., as described in procedures).

The process of sub-goal decomposition continues until an appropriate set of end operations is reached or the level of granularity required for the analysis is achieved. Note that the bottom level of any branch in an HTA should always be a series of operations. Whereas everything above an operation specifies goals, operations state what needs to be done to achieve goals. Thus, operations are the actions to be made by the operator (whether human or non-human). For the supplement use by athletes case study, sub-goal 2 'Research supplements' was decomposed as follows:

2 Research supplements
  2.1 Check banned list
  2.2 Check supplements
  2.3 Check batch testing

As noted above, decomposition should continue until an appropriate set of end operations is reached or the level of granularity required for the analysis is achieved. For example, from the decomposition above the sub-goal 'Check supplements' was decomposed as follows:

2.2.1 Check contents
2.2.2 Check country of production
2.2.3 Check safety guidelines
2.2.4 Check compliance with Therapeutic Goods Association registration
2.2.5 Check Therapeutic Use Exemption requirements
2.2.6 Check specific education materials

Step 7: Plans analysis

Once all sub-goals have been decomposed and described, the plans should be developed. As described earlier, plans dictate the sequencing of goals, sub-goals, and operations (Salmon et al., 2022). It is important here to ensure that the plans include all of the context conditions and triggers that dictate the order with which sub-goals and operations are actually carried out. For example, a simple linear plan would say 'Do 1, then 2, and then 3, then EXIT'; however, if the sub-goals are undertaken in parallel the plan would say 'Do 1 then 2 and 3 then EXIT'. Various forms of plan exist, and the method is flexible in that new types of plans can be added should the analysis require it. An overview of different plan types is presented in Table 3.1.

For the supplement use by athletes example, the overall plan for 'Optimise health and performance through nutritional supplement use' was defined as:

Plan. Do 1 then 2, If TUE is required then do 3 then 4 then 5 then 6 then EXIT, If TUE is not required then do 4 then 5 then 6 then EXIT.

TABLE 3.1 HTA plan types (Salmon et al., 2022)

Plan	*Example*
Linear	Do 1, then do 2, then do 3 then EXIT
Non-linear	Do 1, 2 and 3 in any order then EXIT
Simultaneous	Do 1, then 2 and 3 together then EXIT
Branching	Do 1, if x is present, then do 2 and 3 then EXIT, if x is not present then do 4 then EXIT
Cyclical	Do 1, then do 2, then do 3 and repeat until X then EXIT
Selection	Do 1, then do 2 or 3 as required

Step 8: Construct HTA diagram

Once all of the goals, sub-goals, operations, and plans are described, the next step involves creating the HTA diagram. The HTA is best expressed as a hierarchical tree diagram (see Figure 3.2) but it can also be presented in a tabular format. Various HTA software tools are available to support this process; however, drawing packages such as Microsoft Visio can also be used.

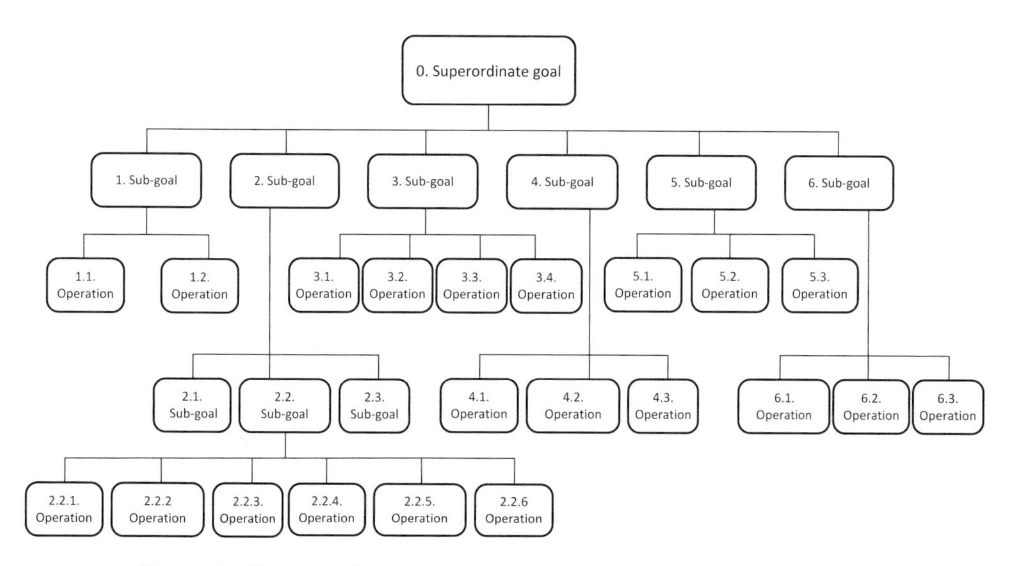

FIGURE 3.2 Example HTA tree diagram.

Step 9: SME review

Once the initial draft HTA is complete, it should be reviewed by as many SMEs as possible within the project constraints. This involves taking SMEs through the HTA and querying whether there are missing sub-goals or operations, whether the plans reflect real life, and whether the terminology is correct throughout. The SME feedback should then be used to refine the HTA with the revised version subjected to further review until the SMEs agree that it provides a valid representation of the task or system under analysis. It is normal practice for the HTA to go through multiple iterations before it is complete (Salmon et al., 2022).

Step 10: Conduct additional analyses using HTA extension methods

A strength of HTA is that the outputs act as the input to various additional analysis methods. For example, error identification methods such as the Systematic Human Error Reduction and Prediction Approach (SHERPA) (Embrey, 1986) and more recently the Net-HARMS (Dallat et al., 2017) risk assessment method require a HTA description as their primary data input. This enables HTA to be applied for various purposes, including interface design and evaluation, job design, training programme design and evaluation, error prediction and risk assessment, team task analysis, situation awareness requirements analysis, allocation of functions analysis, workload assessment, and procedure design (Stanton, 2006; Stanton et al., 2013).

Further information and guidance on HTA can be found in Stanton et al. (2013) and Salmon et al. (2022).

Advantages

- HTA requires minimal training and is easy to apply.
- The outputs are informative and act as the input for various additional analyses, such as error identification, training needs analysis, interface design and evaluation, and allocation of function analysis.
- HTA is a highly flexible technique that can be applied in any domain for a variety of purposes.
- The output provides a comprehensive description of the task or system under analysis.
- Conducting the HTA gives the analyst considerable insight into the task or system under analysis.
- Tasks can be analysed to any required level of detail, depending on the aims of the analysis and resources available.

Disadvantages

- HTA provides mainly descriptive information rather than analytical information.
- Conducting HTA for large, complex tasks or systems can be laborious and time-consuming.
- The initial data collection phase is time-consuming and requires analysts to be competent in a variety of methods, such as interviews, observations, and questionnaires.
- Reliability can be a concern in some instances.
- Applying HTA has been labelled as more of an art than a science (Stanton et al., 2013) and extensive practice is required before analysts become proficient.

Related methods

As noted, HTA often forms the first step in various analyses, such as human error identification, human reliability analysis, training needs analysis, allocation of functions analysis, and workload assessment. Stanton (2006) outlines a range of additional analysis methods, including interface design, error prediction, workload analysis, team performance assessment, and training requirement identification. In sport, methods similar to HTA are often used to break down and analyse complex tasks or coaching strategies. One such methodology is performance analysis, which often employs techniques like notational analysis or video analysis to study the specific actions, movements, and strategies in a game or sporting event. Another related concept is tactical periodisation, particularly prevalent in team sports. This method systematically plans training tasks focused on tactical development within the game model, allowing a hierarchical approach to improving individual and team performance. Skill acquisition models are also sometimes used in sports research to break down the stages of learning a new skill, often employing video and statistical analysis to evaluate performance at various stages.

Approximate training and application times

HTA often requires significant training before analysts become proficient in the method. Patrick, Gregov, and Halliday (2000) reported unsatisfactory results after providing a few hours of training to students for a simple HTA task. Stanton and Young (1999) report that the training and application time for HTA is substantial. A survey by Ainsworth and Marshall (1998) found that more experienced practitioners produced more complete and acceptable analyses.

The application time is dependent upon the size and complexity of the task or system under analysis. For large, complex tasks, the application time for HTA would be high. Salmon et al. (2010) report that HTA application times are high, stating that the requirement for a detailed HTA can increase application times to almost double those of other methods such as CWA (see Chapter 4).

Reliability and validity

When testing the reliability and validity of various methods, Stanton and Young (1999) reported that HTA achieved an acceptable level of validity but poor level of reliability. According to Stanton et al. (2013) analysts with differing levels of experience will likely produce different analyses of the same task (inter-rater reliability) and that the same analyst may produce different analyses on different occasions for the same task (intra-analyst reliability).

Tools needed

Initial drafts of a HTA can be developed using pen and paper, whiteboards, post-it notes, or Notepad. The final HTA output can be developed and presented in a number of software applications, such as Microsoft Visio, Microsoft Word, and Microsoft Excel. A number of dedicated HTA software tools also exist, such as the HTA tool (Salmon et al., 2010).

Case study example: decomposing nutritional and dietary supplement use by athletes to avoid unintentional doping

Background

Unintentional doping through nutritional and dietary supplement use is an intractable yet preventable issue in elite sport. Research has demonstrated that athletes regularly use nutritional and dietary supplements, and in some sports, it is estimated that all athletes use supplements (Erdman et al., 2007; Nieper, 2005; Tscholl et al., 2010). In addition, research has indicated that a considerable percentage of supplements may contain prohibited substances, be mislabelled or inadequately labelled, or even specifically 'spiked' with prohibited substances (Baylis et al., 2001; de Hon & Coumans, 2007; De Cock et al., 2001; Delbeke et al., 2002; Duiven et al., 2021; Geyer et al., 2011; Maughan, 2005; Van der Merwe & Grobbelaar, 2005). In one study, an assessment of 216 sports nutrition supplements claiming to modulate hormone regulation, stimulate muscle mass gain, increase fat loss, and/or boost energy were analysed, found that 38% contained undeclared banned substances (Duiven et al., 2021). As such, athletes can be unwittingly and unintentionally exposed to doping when consuming supplements, as they are oblivious to the specific ingredient content (Chan et al., 2016; Chan et al., 2019).

The high rate of supplement usage among athletes, coupled with potential of contaminated and mislabelled supplements, has resulted in an ongoing problem of athletes returning adverse analytical findings through supplement use (Outram & Stewart, 2015). Alarmingly, it is estimated that up to 9% of all positive doping tests are caused by athletes using poorly labelled sports nutrition and dietary supplements (Outram & Stewart, 2015). Further, a WADA study into athlete doping has indicated that nutritional and dietary supplements are the most important risk factor for unintentional doping (WADA, 2022). While the risks associated with acquiring contaminated supplements may be reduced at elite levels through enhanced controls over supplement sourcing (Outram & Stewart, 2015), up to 50% of elite athletes purchase supplements through standard retail sources such as stores and the internet (Striegel et al., 2005).

Given the inherent risk associated with athletes and supplement use, the aim of this case study was to develop a HTA to demonstrate the behaviour and actions required for safe and compliant nutritional supplement use by athletes.

Method

For this case study, data collection activities included relevant peer-reviewed literature, and documentation review (e.g., World Anti-Doping Code) and education materials on supplement use and unintentional doping in sport (e.g., National Anti-Doping Agencies). The authors developed the HTA during an in-person workshop.

Results

See Figure 3.3 for the developed HTA of optimising athlete health and performance through nutritional and dietary supplement use.

Discussion

The developed HTA demonstrates the complex series of goals, sub-goals, and operations required by athletes seeking to use nutritional and dietary supplements to optimise health and performance. Given the numerous sub-goals and operations, it is understandable that athletes are susceptible to unintentional doping through supplement use. Decomposing this inherent complexity through a HTA can help to identify potential areas for improvements in the process. For example, the HTA highlights that within the current system there are limited opportunities for intervention from outside of the athlete. Only the discussion with coaches and support personnel, and obtaining a TUE represent potential intervention points to assist with athlete decision-making and actions. However, athletes are solely responsible for undertaking these steps, and are not required to complete them prior to using supplements. The HTA itself could be used as an educational resource for athletes, as it steps out the required process to avoid unintentional doping. The HTA could also be supplemented by including the appropriate anti-doping contacts and educational resources at the various identified sub-goals and operations. As noted

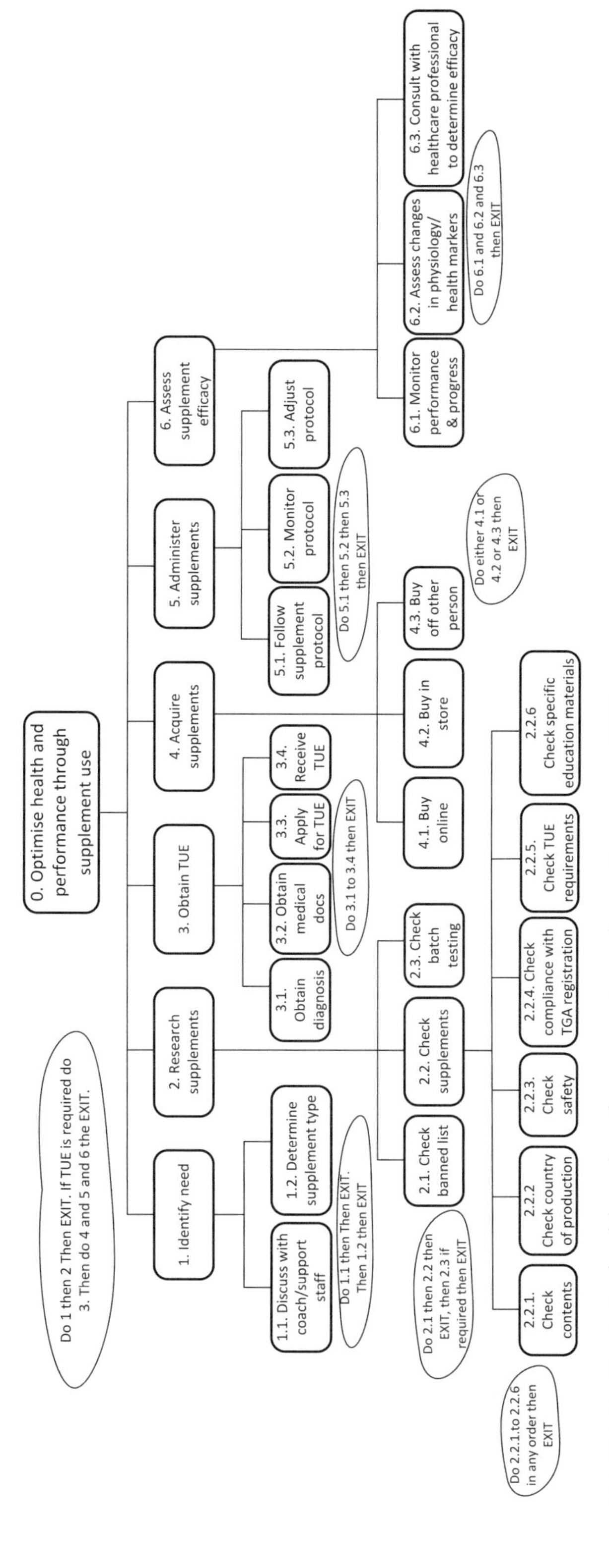

FIGURE 3.3 Optimise health and performance through supplement use HTA.

earlier, the HTA could also be used on conjunction with error identification or prospective risk assessment methods to determine where vulnerabilities exist within the system leading athletes to engage in unintentional doping.

While the current case study was used to identify potential areas of risk for athletes, there are numerous other applications of HTA to optimise sport research and practice. For example, identifying areas of improvement for athlete and team performance; training design; analysing skill acquisition and learning; recovery and rehabilitation processes; evaluation of coaching strategy; and human and technology interaction, among others.

Recommended Reading

Annett, J., et al. (1971). *Task analysis*. Department of Employment Training Information Paper 6. London: HMSO.

Annett, J. (2004). Hierarchical task analysis. In D. Diaper & N. A. Stanton (Eds.), *The handbook of task analysis for human-computer interaction* (pp. 67–82). Mahwah, NJ: Lawrence Erlbaum Associates.

Salmon, P. M., Stanton, N. A., Jenkins, D. P., & Walker, G. H. (2010). Hierarchical task analysis versus cognitive work analysis: Comparison of theory, methodology, and contribution to system design. *Theoretical Issues in Ergonomics Science*, 11(6), 504–531.

Stanton, N. A. (2006). Hierarchical task analysis: Developments, applications, and extensions. *Applied Ergonomics*, 37(1), 55–79.

Stanton, N. A., Salmon, P. M., Rafferty, L., Walker, G., Baber, C., & Jenkins, D. P. (2013). *Human factors methods: A practical guide for engineering and design*. Second Edition. Aldershot: Ashgate.

References

Ainsworth, L., & Marshall, E. (1998). Issues of quality and practicability in task analysis: Preliminary results from two surveys. *Ergonomics*, 41(11), 1607–1617.

Annett, J. (2004). Hierarchical task analysis. In D. Diaper & N. A. Stanton (Eds.), *The handbook of task analysis for human-computer interaction* (pp. 67–82). Mahwah, NJ: Lawrence Erlbaum Associates.

Annett, J., Duncan, K. D., Stammers, R. B., & Gray, M. J. (1971). *Task analysis*. London: Her Majesty's Stationery Office.

Baylis, A., Cameron-Smith, D., & Burke, LM. (2001). Inadvertent doping through supplement use by athletes: Assessment and management of the risk in Australia. *International Journal of Sport Nutrition and Exercise Metabolism*, 11, 365–383.

Bedinger, M., Walker, G. H., Piecyk, M., Greening, P., & Krupenia, S. (2015). A hierarchical task analysis of commercial distribution driving in the UK. *Procedia Manufacturing*, 3, 2862–2866.

Chan, D. K. C., Ntoumanis, N., Gucciardi, D. F., Donovan, R. J., Dimmock, J. A., Hardcastle, S. J., & Hagger, M. S. (2016). What if it really was an accident? The psychology of unintentional doping. *British Journal of Sports Medicine*, 50(15), 898–899.

Chan, D. K. C., Tang, T. C. W., Yung, P. S. H., Gucciardi, D. F., & Hagger, M. S. (2019). Is unintentional doping real, or just an excuse? *British Journal of Sports Medicine*, 53(15), 978–979.

Dallat, C., Salmon, P. M., & Goode, N. (2018). Identifying risks and emergent risks across sociotechnical systems: The NETworked hazard analysis and risk management system (NET-HARMS). *Theoretical Issues in Ergonomics Science*, 19(4), 456–482.

De Cock, K., Delbeke, F., Van Eenoo, P., Desmet, N., Roels, K., & De Backer, P. (2001). Detection and determination of anabolic steroids in nutritional supplements. *Journal of Pharmaceutical and Biomedical Analysis*, 25(5), 843–852.

de Hon, O., & Coumans, B. (2007). The continuing story of nutritional supplements and doping infractions. *British Journal of Sports Medicine*, 41(11), 800–805.

Delbeke, F.T., Van Eenoo, P., Van Thuyne, W., & Desmet, N. (2002). Prohormones and sport. *The Journal of Steroid Biochemistry and Molecular Biology*, 83(1), 245–251.

Demirel, D., Butler, K.L., Halic, T., Sankaranarayanan, G., Spindler, D., Cao, C., Petrusa, E., Molina, M., Jones, D.B., De, S., & deMoya, M.A. (2016). A hierarchical task analysis of cricothyroidotomy procedure for a virtual airway skills trainer simulator. *The American Journal of Surgery*, 212(3), 475–484.

Duiven, E., van Loon, L. J., Spruijt, L., Koert, W., & de Hon, O. M. (2021). Undeclared doping substances are highly prevalent in commercial sports nutrition supplements. *Journal of Sports Science & Medicine*, 20(2), 328.

Embrey, D. E. (1986). SHERPA: A systematic human error reduction and prediction approach. In *Proceedings of the international topical meeting on advances in human factors in nuclear power systems*. Knoxville, Tennessee.

Erdman, K. A., Fung, T. S., Doyle-Baker, P. K., Verhoef, M. J., & Reimer, R. A. (2007). Dietary supplementation of high-performance Canadian athletes by age and gender. *Clinical Journal of Sport Medicine*, 17(6), 458–464.

Fargnoli, M., Lombardi, M., & Puri, D. (2019). Applying hierarchical task analysis to depict human safety errors during pesticide use in vineyard cultivation. *Agriculture*, 9(7), 158.

Geyer, H., Braun, H., Burke, L. M., Stear, S. J., & Castell, L. M. (2011). A–Z of nutritional supplements: Dietary supplements, sports nutrition foods and ergogenic aids for health and performance—Part 22. *British Journal of Sports Medicine*, 45(9), 752–754.

Kirwan, B., & Ainsworth, L. K. (Eds.). (1992). *A guide to task analysis: The task analysis working group*. London, CRC Press.

Maughan, R. J. (2005). Contamination of dietary supplements and positive drug tests in sport. *Journal of Sports Sciences*, 23(9), 883–889.

Mills, S. (2007). Contextualising design: Aspects of using usability context analysis and hierarchical task analysis for software design. *Behaviour & Information Technology*, 26(6), 499–506.

Naweed, A., Balakrishnan, G., & Dorrian, J. (2018). Going solo: Hierarchical task analysis of the second driver in "two-up" (multi-person) freight rail operations. *Applied Ergonomics*, 70, 202–231.

Nieper, A. (2005). Nutritional supplement practices in UK junior national track and field athletes. *British Journal of Sports Medicine*, 39(9), 645–649.

Nixon, J., & Braithwaite, G. R. (2018). What do aircraft accident investigators do and what makes them good at it? Developing a competency framework for investigators using grounded theory. *Safety Science*, 103, 153–161.

Outram, S., & Stewart, B. (2015). Doping through supplement use: A review of the available empirical data. *International Journal of Sport Nutrition & Exercise Metabolism*, 25(1), 54–59.

Patrick, J., Gregov, A., & Halliday, P. (2000). Analysing and training task analysis. *Instructional Science*, 28, 51–79.

Ramos, M. A., Utne, I. B., & Mosleh, A. (2019). Collision avoidance on maritime autonomous surface ships: Operators' tasks and human failure events. *Safety Science*, 116, 33–44.

Salmon, P. M., Stanton, N. A., Jenkins, D. P., Walker, G. H. (2010). Hierarchical task analysis versus cognitive work analysis: Comparison of theory, methodology, and contribution to system design. *Theoretical Issues in Ergonomics Science*, 11(6), 504–531.

Salmon, P. M., Stanton, N. A., Walker, G. H., Hulme, A., Goode, N., Thompson, J., & Read, G. J. M. (2022). *Handbook of Systems Thinking Methods*. Boca Raton, FL, CRC Press.

Smith, P. A., Baber, C., Hunter, J., & Butler, M. (2008). Measuring team skills in crime scene investigation: Exploring ad hoc teams. *Ergonomics*, 51(10), 1463–1488.

Stanton, N. A., & Young, M. S. (1999). What price ergonomics? *Nature*, 399(6733), 197–198.

Stanton, N. A., Salmon, P. M., Walker, G. H., & Jenkins, D. P. (2010). Is situation awareness all in the mind? *Theoretical Issues in Ergonomics Science*, 11(1–2), 29–40.

Stanton, N. A., Salmon, P. M., & Rafferty, L. A. (2013). *Human factors methods: A practical guide for engineering and design*. Aldershot, Ashgate.

Stanton, N. A. (2006). Hierarchical task analysis: Developments, applications, and extensions. *Applied Ergonomics*, 37(1), 55–79.

Striegel, H., Vollkommer, G., Horstmann, T., & Niess, A. M. (2005). Contaminated nutritional supplements–legal protection for elite athletes who tested positive: A case report from Germany. *Journal of Sports Sciences*, 23(7), 723–726.

Tscholl, P., Alonso, J.M., Dollé, G., Junge, A., & Dvorak, J. (2010). The use of drugs and nutritional supplements in top-level track and field athletes. *American Journal of Sports Medicine*, 38(1), 133–140.

van der Merwe, P. J. V., & Grobbelaar, E. (2005). Unintentional doping through the use of contaminated nutritional supplements. *South African Medical Journal*, 95(7), 510–511.

World Anti-Doping Agency. (2022). Athlete vulnerabilities research project. Descriptive report on sport stakeholders' beliefs about athlete doping vulnerabilities and related factors. Available from: www.wada-ama.org/en/news/report-key-areas-athlete-vulnerabilities-doping

Yu, D., Minter, R. M., Armstrong, T. J., Frischknecht, A. C., Green, C., & Kasten, S. J. (2014). Identification of technique variations among microvascular surgeons and cases using hierarchical task analysis. *Ergonomics*, 57(2), 219–235.

4

COGNITIVE WORK ANALYSIS

Background

Sports are more than athleticism and tactics, they are underpinned by intricate and complex systems that govern everything from individual plays to entire seasons, and from individual players through to entire leagues. This complexity can manifest in the visible tactics of a team, or the less obvious operational aspects of entire sporting organisations. Unfortunately, traditional analysis methods in sports often overlook this complexity, opting to focus on isolated components rather than their interactions within the system (Mclean et al., 2017). To cope with complexity, including the various pressures on individuals, teams, sports organisations and leagues, adaptation is often required.

CWA is a sociotechnical systems analysis and design framework that has been used extensively for understanding the structure and behaviour of complex systems (Bisantz & Burns, 2008; Stanton et al., 2017). The CWA framework provides a series of phases and associated analytical methods that focus on identifying the constraints on system behaviour and the resulting possibilities for behaviour and adaptation. This allows analysts to understand what constraints exist, what their impact on system functioning, and how constraints can be modified to improve system performance. The framework is formative which means that it considers the possibilities for behaviour within the set of system constraints, as opposed to analysing the behaviours that should occur (as provided by normative analysis methods) or what behaviour actually occurs (as provided by descriptive analysis methods). The formative nature of the approach allows CWA to be used for both analysis and design, with formative applications identifying possibilities to support flexible and adaptive behaviour and supporting the analysis of future 'first-of-their-kind' systems (Naikar et al., 2003; Salmon et al., 2024; Salmon et al., 2022).

The CWA framework comprises five phases, each being used to model behaviour from differing perspectives: work domain analysis (WDA); control task analysis (ConTA); strategies analysis (StrA); social organisation and cooperation analysis (SOCA); and worker competencies analysis (WCA). An overview of each of the phases is provided below with specific emphasis on WDA and SOCA, the two phases used in the case study

DOI: 10.4324/9781003259473-6

at the end of this chapter – the redesign of an Australian Rules Football (AFL) club's football department.

Applications in sport

Specific CWA phases, and combinations of CWA phases have been applied in multiple sports contexts. These applications include modelling overall performance in football (soccer) and elite netball (McLean et al., 2017; 2019), identifying organisational functions and constraints influencing performance in elite netball (Hulme et al., 2019), analysing talent identification and development in soccer (Berber et al., 2020), design of an optimal para-sport development programme (McLean et al., 2021), assessing the perceptions of goalkeeping coaches and players (Berber, 2021), and modelling the functioning of an AFL football club's performance department to support redesign (McLean et al., 2021).

Procedure and advice

A flowchart depicting the CWA procedure is presented in Figure 4.1 (adapted from Salmon et al., 2022). This chapter refers to the commonly applied CWA phases and methods outlined by Vicente (1999), Jenkins and colleagues (2009), and Stanton and colleagues (2017). Further descriptions can be found in other texts (e.g., Vicente, 1999; Jenkins et al., 2009; Naikar, 2013; Salmon et al., 2022).

Step 1: Determine the aims and objectives of the analysis

The first step in applying CWA involves defining the analysis aims and boundaries. The exact aims of the analysis and the boundaries of the system under consideration should be clearly specified to guide the CWA application, including the data collection methods to be used and the phases to be applied. Consideration of scope is important to avoid being too narrow and thus not taking an appropriate systems thinking lens, while also avoiding such a broad scope that the analysis becomes unmanageable. Given that CWA is an iterative framework, the scope and boundary of analysis may be re-considered as the analysis progresses.

Step 2: Select appropriate CWA phases and methods

Once the aims and boundaries are defined, the analysis team select the most appropriate CWA phase(s) and methods to be employed during the analysis. For example, when using the framework for the design of a new device (e.g., athlete tracking technology), it may be that only the first phase, WDA, is required to model the device and how it will integrate into the broader system (McLean et al., 2017; Salmon et al., 2021). Alternatively, when seeking to specify an optimal allocation of functions across roles within an organisation, WDA and SOCA will be useful together (Hulme et al., 2018; McLean et al., 2021). In general, it is recommended that WDA be applied as a starting point as it provides a holistic view of the system. Based on the selection of phases and methods, Steps 4–9 are then conducted as appropriate.

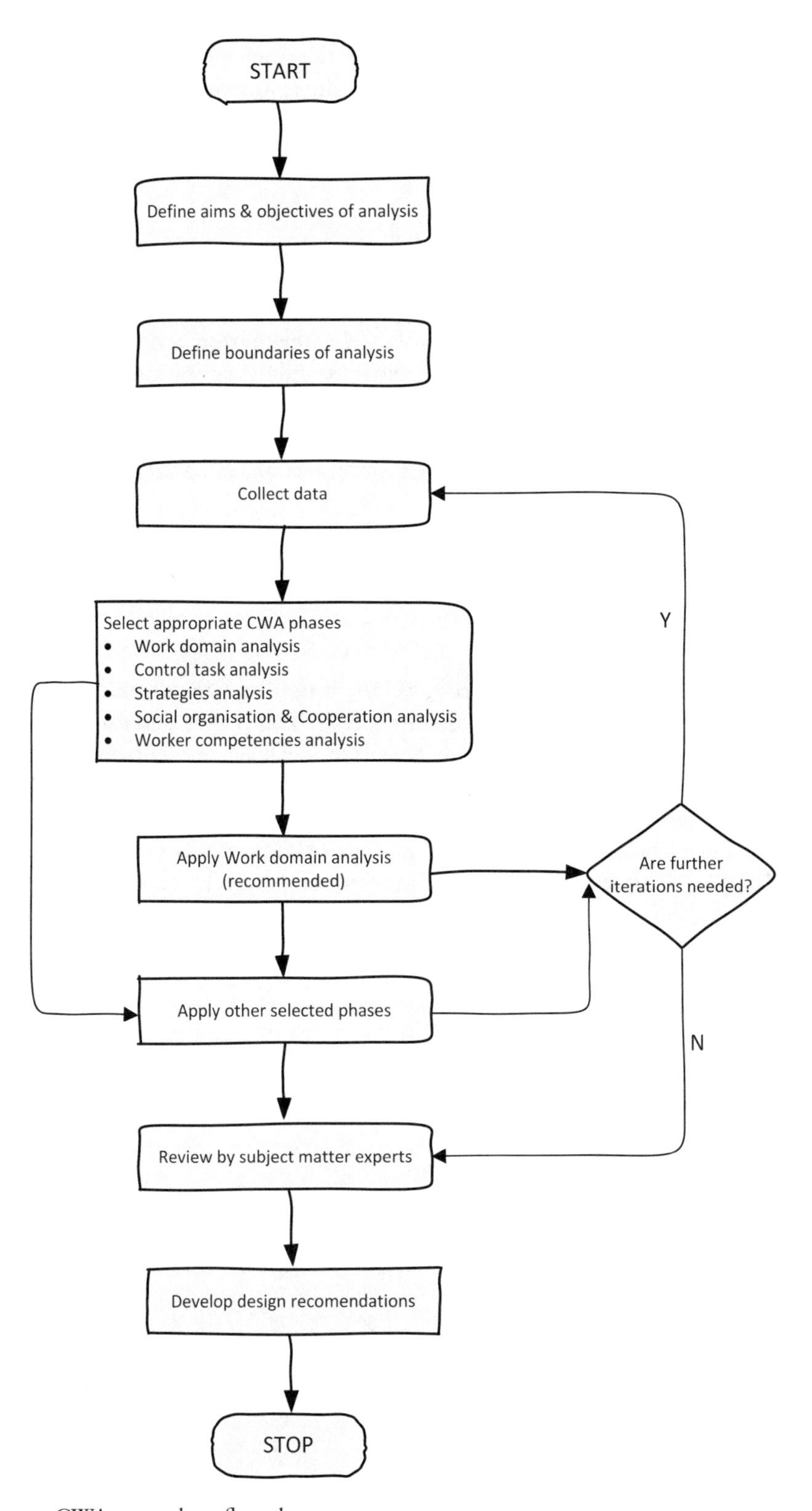

FIGURE 4.1 CWA procedure flowchart.

In the case study example in this chapter, as the aims included both an analysis of football department structure and of how purposes, values, functions, processes, and objects are allocated across different stakeholders, the WDA and SOCA phases were selected.

Step 3: Data collection

Once the aims of the analysis are clearly defined and the appropriate phases and methods are chosen, the next step involves collecting targeted data about the system and its behaviour. The specific data collected is dependent on the phases being applied; however, data collection for CWA typically involves documentation review (e.g., game models, match tactics, club strategy, incident reports, standard operating procedures), SME workshops, direct observations, task walkthroughs, concurrent verbal protocols, structured or semi-structured interviews, e.g., the Critical Decision Method (CDM; O'Hare et al., 2000), questionnaires, and surveys. See Stanton et al., 2013 for additional guidance on these data collection methods.

Recommended data sources for each phase are presented in Table 4.1; however, it should be noted that these are not minimum requirements.

Step 4: Work Domain Analysis

The initial phase of the CWA framework, WDA, is used to provide an event and actor-independent description of the system under analysis. This means that the WDA does not model individuals or their behaviour, nor does it model specific events or situations. For the case study later in this chapter, we depict the functioning of a current AFL club's football department 'system'. The aim of WDA is to describe the purposes of the system and the ecological constraints imposed on the actions of those performing activities within it (Vicente, 1999). This involves using the abstraction hierarchy method to describe the system across five levels of abstraction (Table 4.2).

The abstraction hierarchy (Rasmussen 1987; Vicente 1999) consists of five levels of abstraction, ranging from the most abstract level of purposes to the most concrete level of physical form (Figure 4.2). The labels used for each of the levels of the hierarchy can differ, depending on the aims of the analysis.

The top three levels of the abstraction hierarchy identify the overall objectives of the systems and what it can achieve, whereas the bottom two levels concentrate on the

TABLE 4.1 Example data sources for each CWA phase

	WDA	*ConTa*	*StrA*	*SOCA*	*WCA*
Documentation review	X		X	X	
SME workshops	X	X	X	X	X
Direct observation	X	X	X	X	
Cognitive task analysis interview (e.g., CDM)		X	X	X	X
Task walkthrough	X	X	X	X	
Concurrent verbal protocols		X	X	X	X
Questionnaire/Survey	X			X	

TABLE 4.2 WDA descriptions of levels of abstraction

Level of abstraction	*Description*
Functional purpose/s	The overall purposes of the system and the external constraints imposed on its operation
Values and priority measures	The criteria used for measuring progress towards the functional purpose/s
Purpose-related functions	The general functions that are necessary to be undertaken to achieve the functional purposes
Object-related processes	The functional capabilities and limitations of the physical objects within the system that enable the purpose-related functions to be undertaken
Physical objects	The physical objects within the system that are used to undertake the purpose-relation functions

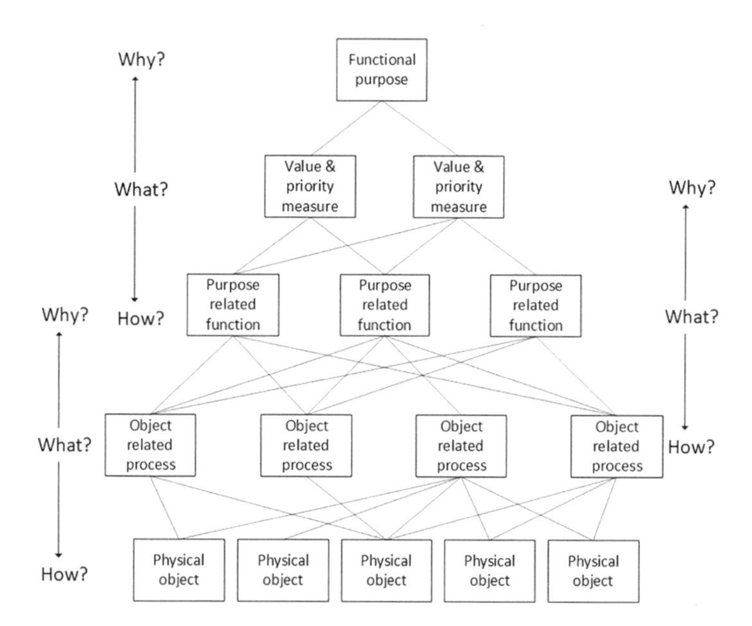

FIGURE 4.2 WDA framework showing the levels of abstraction, and the means-end links 'how-what-why' triad.

physical components and their affordances. The abstraction hierarchy is constructed by considering the system's objectives (top-down) and the system's capabilities (bottom-up).

Nodes at each level of the abstraction hierarchy are connected to nodes at the levels above and below via means-ends links. These are used to show the relationships between objects, processes, functions, values, and functional purposes. As shown in Figure 4.1, for any given node, the linked nodes at the level above describe why that node is required, and the linked nodes at the level below describe how the node is achieved. For example, in a sporting context, the object-related process of 'treatment of player injuries'

TABLE 4.3 Prompts to aid abstraction hierarchy development of an AFL football performance department (adapted from McLean et al., 2021)

Stage	*Question*	*Keywords*	*Examples*
Functional purposes	Why does the football department exist?	Reasons, goals, objectives, aims, intentions, mission	To win games/grand final Player/team development Implement club strategic plan
Values and priority measures	How can we tell whether football department is achieving its purposes?	Criteria, measures, benchmarks	Club reputation Player & team performance Match & season outcomes Staff & player satisfaction Staff & player retention
Purpose-related functions	What functions must be performed by club staff for the football department to achieve its purposes?	Roles, responsibilities, tasks, jobs, occupations, positions, activities, operations	Talent identification & recruitment Performance analysis Coaching and training Load and injury management Manage staff and player health and wellbeing
Object-related processes	What processes are physical objects used to achieve within the football department?	Uses, applications, characteristics, limitations, processes	Data collection & analysis Development of physical strength & athletic capacity Communication
Physical objects	What physical objects are used within the football department?	Tools, equipment, technology, kit, gear, buildings, facilities, infrastructure, people, terrain	Strategic plan Training equipment Gymnasium Finances

is undertaken to achieve the purpose-related function of 'injury prevention, management, and rehabilitation' and involves the use of the physical object 'medical equipment'. This feature of the abstraction hierarchy enables analysts to understand why functions and processes are undertaken, and what resources can be used to undertake them. In Table 4.3, a set of prompts are presented to assist with developing the abstraction hierarchy for an AFL club's football performance department.

Step 5: Control Task Analysis – contextual activity template

During the ConTA phase of CWA, particular attention is given to situation-specific constraints. One method to examine these constraints is through the use of the contextual activity template (CAT), as suggested by Naikar et al. (2006). The CAT serves as a mechanism to map the functions outlined in the abstraction hierarchy across various distinct 'situations'. For example, it can be used to define the functions that are required in pre-season, in-season, and post-season. Similar to the WDA phase, the CAT is actor-independent, meaning it focuses on functions in a broad sense rather than tasks executed by particular human or technological agents.

The initial step in developing a CAT involves defining the situations of interest to model, which can be distinguished by time, location, or a blend of both. Often, it proves beneficial to investigate a variety of types of situations through several CAT diagrams, thereby accommodating diverse analytical objectives.

Step 6: Control Task Analysis – decision ladders

A second method commonly applied within the ConTA phase is the decision ladder (Rasmussen et al. 1994). Decision ladders enable key function-situation cells within the CAT to be explored in terms of the constraints on decision-making. Again, decision ladders are actor-independent and not specific to any single actor; instead, they represent the decision-making process of the combined system. In many cases, the decision-making process may be collaborative, distributed between a range of human and non-human agents.

The decision ladder is used to represent the processes associated with observing and interpreting the current system state, and also represents goal selection, and the planning and execution of tasks and procedures to achieve a target system state (see Stanton et al., 2017). It can be used to describe the decision-making of novices (novices pass through the decision ladder in a linear manner) as well as experts (experts take 'short cuts' through the decision ladder).

Decision ladders are typically populated from interview data, such as the CDM (see O'Hare et al., 2000). In its raw form, decision ladder models provide a list of the information requirements for making a decision. At this stage of the analysis, the relative importance of these information elements is not considered. Amongst other things, the outputs can be useful in supporting the design of training and decision support systems.

Step 7: Strategies Analysis

The StrA phase of CWA identifies the constraints influencing the way in which activity can be conducted. In keeping with the remainder of the framework, it introduces additional detail to the analyses developed in the previous phases. The aim of this phase is to describe the constraints that dictate how a system can be (rather than how it should be or currently is) moved from one state to another. This is commonly achieved by using information flow maps, which identify relevant start state and end states for an activity and all of the strategies that can move the system between those two states. Although information flow maps are typically used, other tools, such as the Strategies Analysis Diagram (Cornelissen et al., 2013) can also be applied. Overall the StrA phase can be particularly useful for exploring flexibility and adaptation within a system (Jenkins et al., 2009).

In the context of sport, the StrA phase could be used to examine the methods and tactics used by athletes, coaches, and teams for specific conditions (e.g., with different opposition team strategies). This analysis could include all the strategies associated with possible techniques or training protocols.

Step 8: Social Organisation and Cooperation Analysis

The SOCA phase of CWA involves identifying how activity is and can be distributed between human and non-human agents within the system. This phase of CWA builds directly upon the outputs of previous phases, such as the abstraction hierarchy, decision ladders, and information flow maps. Conducting the SOCA involves first identifying relevant human and technological actors and then using shading or annotation to analyse the abstraction hierarchy (SOCA-WDA – shows which actors undertake which functions/provide which affordances), CAT (SOCA-CAT – shows which actors conduct which function within each situation), decision ladders (SOCA-DL – shows which actors undertake which processes) and/or information flow maps (SOCA-StrA – shows which actors use which strategies). Note that SOCA can also be used to consider what activities actors could perform given design modifications. Overall, the SOCA provides a systemic description that can be used to explore the optimal allocation of functions across the system, as well as identifying communication and collaboration requirements.

As an example, to apply SOCA to the abstraction hierarchy as part of a SOCA-WDA, the first step involves identifying all the relevant actors within the system and allocating a colour to each. This includes human (e.g., athletes, coaches, high-performance staff) and non-human actors (e.g., GPS, playing equipment, coaching equipment). Next, the analyst works through each of the nodes in the abstraction hierarchy and determines which actors currently perform or provide each node. Depending on the purpose of the analysis, additional SOCA-WDA representations may be created to explore which actors could perform or provide each node, given system design modifications.

Step 9: Worker competencies analysis

The final phase of CWA involves identifying the cognitive competencies required for task performance. The WCA phase uses Rasmussen's (1983) skills, rules, and knowledge framework to classify the cognitive activities employed by actors during task performance. The skills, rules, and knowledge taxonomy (Rasmussen, 1983) enables the analyst to map, for each critical strategy, how the system supports each level of cognitive processing.

In the skills, rules, and knowledge taxonomy, skill-based behaviour is associated with sensory-motor performance which occurs in skilled activity without the requirement for conscious control (e.g., an experienced goalkeeper taking a save). Rule-based behaviour refers to the application of stored rules, based on past experience, to determine behaviour (e.g., a player applying tactics based on their knowledge of opposition team weaknesses). Finally, knowledge-based behaviour is engaged in unfamiliar situations where it is not possible to draw upon past experience and the actor must engage in reasoning to understand the situation and select an appropriate course of action (e.g., responding to a novel and unfamiliar tactic being used by an opposition team).

CWA summary

One of the primary strengths of CWA is that its five-phase analysis approach allows it to be applied to address a wide range of analysis aims and objectives. This flexibility allows for various applications across different sport contexts and for different purposes. To provide a simple example, Table 4.4 illustrates the utility of CWA to support performance analysis of a corner kick in football across the five phases of the CWA framework. Similar analyses could be undertaken for different contexts across sports, e.g., from in-game actions through to entire league governance.

TABLE 4.4 Summary of the CWA phases, functions, and specific examples for analysing a corner kick in football

CWA phase	*Phase function*	*Potential utility in performance analysis of a corner kick*
Work domain analysis (WDA)	Models the system to identify the constraints, independent of specific actors and events.	Identify the interacting components of a corner kick, ball, goal frame, along with their affordances, the different functions that they can achieve, and the range of tactical purposes that a corner kick can fulfil.
Control task analysis (ConTA)	Analyses activity in work domain, and decision-making including information requirements and factors influencing decision-making.	Assess decision-making performance in depth including the information required for different tactics both defensive and attacking player decision-making, e.g., decision-making processes leading to successful and unsuccessful outcomes of a corner kick.
Strategies analysis (StrA)	Identifies the strategies adopted by actors (e.g., players, coaches) to perform tasks in the system. This includes all possible strategies for key tasks.	Identify all of the possible ways a goal could be scored from a corner, e.g., short corner, flick on at the near post.
Social organisation & cooperation analysis (SOCA)	Identifies the division of tasks between human and non-human agents, to determine how the system can work together to enhance performance.	Identifies who does or could perform specific tasks during the corner kick, e.g., which players will attack the ball and which will make decoy runs.
Worker competencies analysis (WCA)	Identifies the cognitive competencies that are required to perform key tasks.	Used with players during post-match review to identify levels of cognition and whether this was appropriate within the situation (e.g., over-reliance on skill-based cognition, meaning that novel information was not integrated to inform course of action chosen).

Advantages

- CWA is a highly flexible framework that can be applied in any domain for a variety of purposes.
- The output provides a comprehensive description of the constraints impacting behaviour within the system under analysis.
- Creating the CWA outputs gives the analyst considerable insight into the system under analysis.
- The method is well established, with many examples available in the published literature.

Disadvantages

- Conducting CWA analyses can be laborious and time-consuming, especially if all phases are used.
- The initial data collection phase is time-consuming and requires analysts to be competent in a variety of methods, such as interviews, observations, and questionnaires.
- Initial training time can be a barrier to uptake of the framework.
- Reliability can be a concern in some instances.

Related methods

Various methods can be used to gather the data required for each phase including cognitive task analysis interviews such as the CDM, concurrent verbal protocols, direct observations, cognitive walkthroughs, and hierarchical task analysis (HTA) (see Chapter 3). The WDA-BN approach is an extension to the first phase of CWA, WDA, that systematically breaks the means-ends relationships between nodes in the WDA abstraction hierarchy to enable analysts to identify strategies to mitigate risks associated with functions not being performed or being poorly performed, or to purposefully disrupt the system (Salmon et al., 2021).

Training and application times

Due to the number of phases and methods within the phases, the training time associated with the CWA framework is high. Similarly, the application time can also be high if all five phases are applied. However, in many cases, analysts choose to use a selection of phases that will be sufficient to answer their aims and objectives. For example, a number of applications involve the WDA phase only, and assuming the analyst has experience with the framework, and has access to the necessary information or domain experts to input to the model, this phase can be completed relatively quickly.

Reliability and validity evidence

The formative nature of the CWA framework makes the assessment of its reliability and validity somewhat difficult. However, some testing has been conducted. For example, Burns and colleagues (2004) conducted a qualitative comparison of two independently created WDA models which described similar military command and control environments. Although the models were developed with different scopes and purposes, they found a reasonable level of consistency in the models, suggesting a level of reliability. In a later study, Cornelissen and colleagues (2013) tested the criterion-referenced validity and test-retest reliability of the Strategies Analysis Diagram method using novice analysts. They reported that the method did not show high levels of reliability and validity. However, when individual results were pooled, the validity was much improved, suggesting that group modelling processes may be superior to individual model development processes. However, as noted above, this study used novice analysts and therefore it may be that more reliable outcomes are gained with higher analyst expertise.

Tools required

No specific tools are required to conduct CWA, however, a software tool has been developed to support users to develop the graphical representations required in CWA, including user guidance to support users through the five phases (Jenkins et al., 2007). If a specialised software tool is not used, the representations can be drawn in any illustration software such as Microsoft Visio™ or they could be hand drawn with key aspects drawn formally in illustration software for presentation purposes.

Case study: using CWA to understand an AFL football club performance department functioning

Background

There is increasing interest in applying complexity and systems thinking-based methods, such as CWA, to understand and optimise sports systems (Hulme et al., 2019; McLean et al., 2017; 2019; Salmon & McLean, 2020). Methods such as CWA are useful as they can be used to describe sports organisations, their key functions, and the constraints that influence performance at the athlete, team, and organisational level. As elite sport becomes more complex, competitive, diverse, and increasingly reliant on technology, the importance of adopting a complex systems thinking approach cannot be understated. This was even more so given the COVID-19 Pandemic and the additional constraints placed on club functioning such as travel restrictions, isolation and physical distancing requirements. A successful return to competition required agility, innovation, and ultimately significant modifications to current operations.

The aim of this case study was to use two phases of the CWA framework (Vicente, 1999), WDA and SOCA, to develop and analyse a complex systems model of an AFL club football department. The intention was to identify potential modifications to the club's operations to support a return to competition following the COVID-19 crisis which severely impacted the sporting industry globally.

Methods

The WDA-SOCA was developed across two workshops with five subject matter experts (SMEs) with extensive experience in the AFL (16.2 ± 6.1 years), across a range of different roles including players, football director, general manager of football, and specialists in strength and conditioning, biomechanics, performance analysis, high-performance

TABLE 4.5 SOCA descriptions for the levels of abstraction

Level of abstraction	*Description*
Functional purpose	Actors who contribute to the functional purpose as part of their work in the football department.
Values and priority measures	Actors who hold the value and priority as part of their work in the football department.
Purpose-related functions	Actors who undertake the function as part of the work in the football department.
Object-related processes	Actors who undertake the object-related process as part of their work in the football department.
Physical objects	Actors who use the physical object as part of their work in the football department.

management, AFL governance, coach innovation and education, football strategy and innovation, and playing list management. Two online workshops were held to first develop the WDA and second to undertake the SOCA phase. The first workshop involved the development of the WDA. During the workshop participants were asked to respond to a set of WDA development questions which were presented in conjunction with relevant keywords and examples (see Table 4.3).

A second workshop was then held to review and refine the WDA and undertake the SOCA phase. After reviewing and refining the WDA, participants were asked to create a list of all actors who held a role in the football department. Once the list of actors was finalised participants were asked to identify which were associated with the functional purposes, values and priority measures, purpose-related functions, object-related processes, and physical objects specified in the WDA (based on the prompts provided in Table 4.5).

Results

Work domain analysis

A summary of the AFL club football department abstraction hierarchy is presented in Figure 4.3. The full model can be found in Mclean et al. (2021) (open access article).

According to the abstraction hierarchy, the football department has two functional purposes: to win premierships and to achieve a sustainably successful and progressively improving football programme.

A total of 25 values and priority measures were identified. These can be broadly grouped into seven categories. The first set includes values relating to player and team performance and development, such as matches won, percentages, maximising player talent, and continual player improvement. The second set includes values relating to player health and wellbeing, such as player physical conditioning, minimising injuries, and maximising player health and wellbeing. The third set includes values relating to club finances, including optimising department spend and optimising player spend (i.e., salary caps). The fourth set includes values relating to staff health and wellbeing. The fifth set of values relate to compliance such as minimising positive drug tests (both illicit drugs and

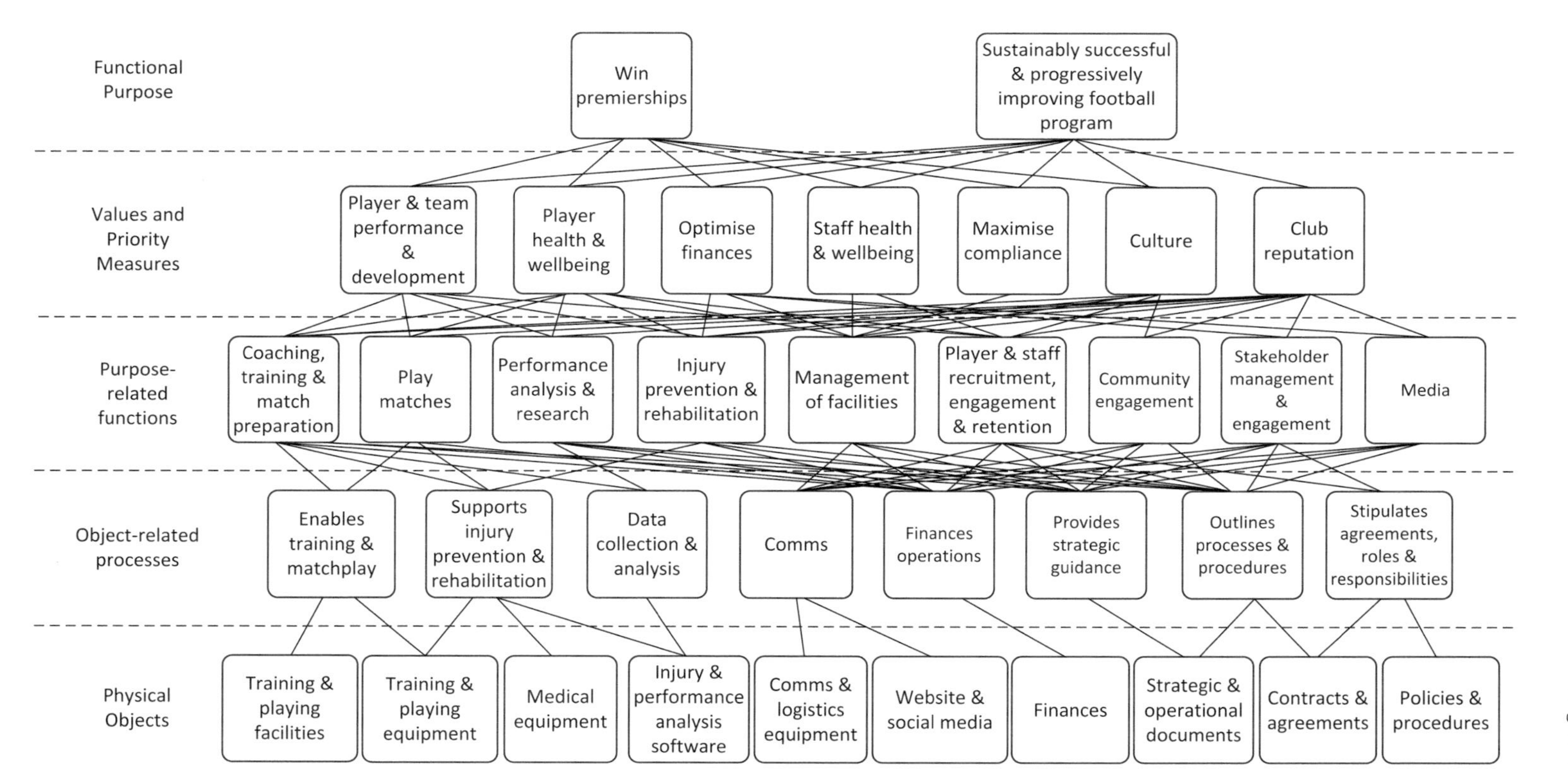

FIGURE 4.3 Summary abstraction hierarchy of the AFL club football department.

Each level contains summary nodes (apart from the functional purposes) that encapsulate more detailed nodes from the overall model. For example, the value and priority measure node of 'player & team performance & development' included 'matches won', 'percentage differential', 'continual player improvement', and 'maximising player talent'.

performance enhancing drugs) and maximising compliance with AFL rules and regulations. The sixth set includes values which relate to the development and maintenance of club culture, such as player inspiration, player and staff engagement, and embracing and supporting diversity in the playing list. Finally, the seventh set includes values which contribute to maintenance of the club's reputation.

A total of 40 purpose-related functions were identified. These include functions relating to coaching, training and match preparation, playing matches, performance analysis and research, injury prevention and rehabilitation, management of facilities, player and staff recruitment, engagement and retention, community engagement, stakeholder management and engagement, and media.

At the bottom level of the abstraction hierarchy, 38 objects were identified, including training and playing facilities and equipment, medical equipment, injury and performance analysis software, communications and logistics equipment, website and social media, finances, strategic and operational documents, contracts and agreements, and policies and procedures. According to the abstraction hierarchy, the physical objects support 28 object-related processes.

Social Organisation and Cooperation Analysis

The AFL club's football department actors identified and considered in the SOCA are presented in Table 4.6.

The results of the SOCA are presented in Figures 4.4–4.7 (Adapted from McLean et al., 2021). The SOCA results demonstrate how functions and processes are distributed

TABLE 4.6 AFL club football department actors (adapted from Mclean et al., 2021)

Code	*Actors*	*Code*	*Actors*
1	Players	30	Player welfare assistant
2	Leadership group (players)	31	Indigenous welfare
3	Football director	32	COO/GM of football
4	Head coach	33	Head of football
5	Coach	34	Assistant to football department
6	Coach	35	Head of women's football
7	Coach	36	Head of list management
8	Development coach	37	National recruiting manager
9	Development coach	38	State recruiting manager
10	Coach	39	Opposition analyst/pro scout
11	Leadership consultant	40	Football Program Advisor (list consultant)
12	Leadership and development coach	41	Psychologist
13	Head of football operations	42	Head trainer
14	Head of performance	43	Podiatrist
15	Head of sport science	44	Yoga/Pilates
16	S & C coaches	45	Lead football analyst
17	S & C rehabilitation	46	Football analyst
18	Development S & C	47	Football analyst
19	GPS analytics	48	Football analyst
20	Doctor	49	Senior analyst
21	Part time Doc	50	Sleep consultant

(*Continued*)

TABLE 4.6 (Continued)

Code	*Actors*	*Code*	*Actors*
22	Part time Doc	51	Legal Counsel and Special Projects
23	Physiotherapist	52	Pastor (part time)
24	Physiotherapist (part time)	53	Massage therapists × 6
25	Physiotherapist (part time)	54	Recruiting administrator
26	Dietician/nutritionist	55	Recruiting scouts × 6
27	Kit man	56	Match day help
28	Facilities manager/staff × 2	57	Analytics interns × 6
29	Player welfare manager	58	Nutrition/S & C interns X 4

Note: Multiple actors that perform the same roles is depicted as × 2, × 4, and × 6. For example, 'Recruiting scouts × 6' indicates that 6 people perform this role.

across the actors within the system. A Figure is not presented for the Functional Purposes level as all actors were deemed to contribute to both 'win premierships' and 'sustainably successful and progressively improving football programme'.

Discussion

The analysis produced multiple insights which are relevant for the optimisation of sports performance departments in general, and were relevant for streamlining football department operations post-COVID-19.

The initial finding of the case study demonstrated the inherent complexity of an AFL football department via the multiple and interacting components that influence the behaviour, and the diverse set of actors who share responsibility for the performance of the system. Further, the current analysis identified strengths of the AFL clubs' football department, as well as potential conflicts between systems components. Several strengths of the football department were identified in the WDA including important functions outside of playing football such as community engagement, staff wellbeing, and the development of culture and club reputation (Jones et al., 2009). The modelling also identified conflicts within the system that will assist the club to redesign operations, and the analysis has important messages for elite sports organisations generally. Firstly, sporting organisations should pursue appropriate goals that reflect the actual state of the system. For example, the functional purpose of 'Win Premiership' may be considered a stretch goal. As an aspirational functional purpose this may be appropriate; however, it is important to note that stretch goals require specific, achievable, attainable, and measurable sub-goals which move the organisation towards the stretch goal. Secondly, the measures used to assess whether the goals of the system are being achieved need to be specific and measurable in order to obtain valid assessments. For example, the extent to which data is available to enable the football department to understand whether they are achieving values and priorities is not clear, and it is questionable whether valid measures exist for some of the values and priorities (e.g., culture). Lastly, the SOCA revealed that there are a large number of actors within the football department. Examination of specific roles is required to determine whether it is feasible to reduce the number of actors while still achieving the functions and values and priorities specific in the abstraction hierarchy. Shifting to a generalist model that combines specialised roles and objects may increase

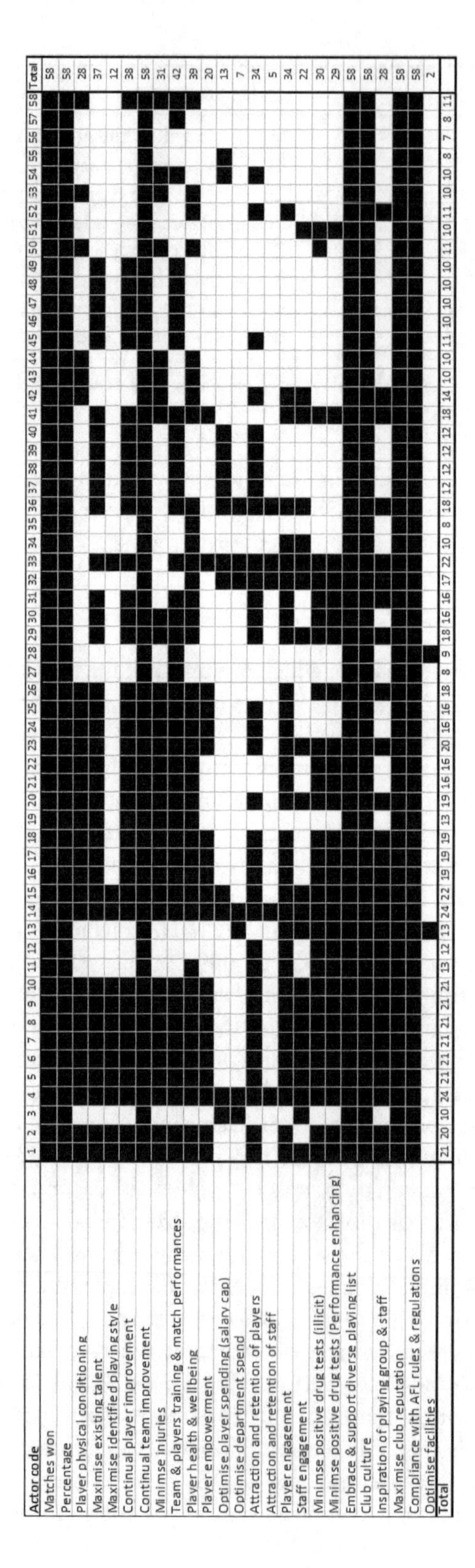

FIGURE 4.4 SOCA showing the football department actors associated with the values and priority measures identified in the abstraction hierarchy.

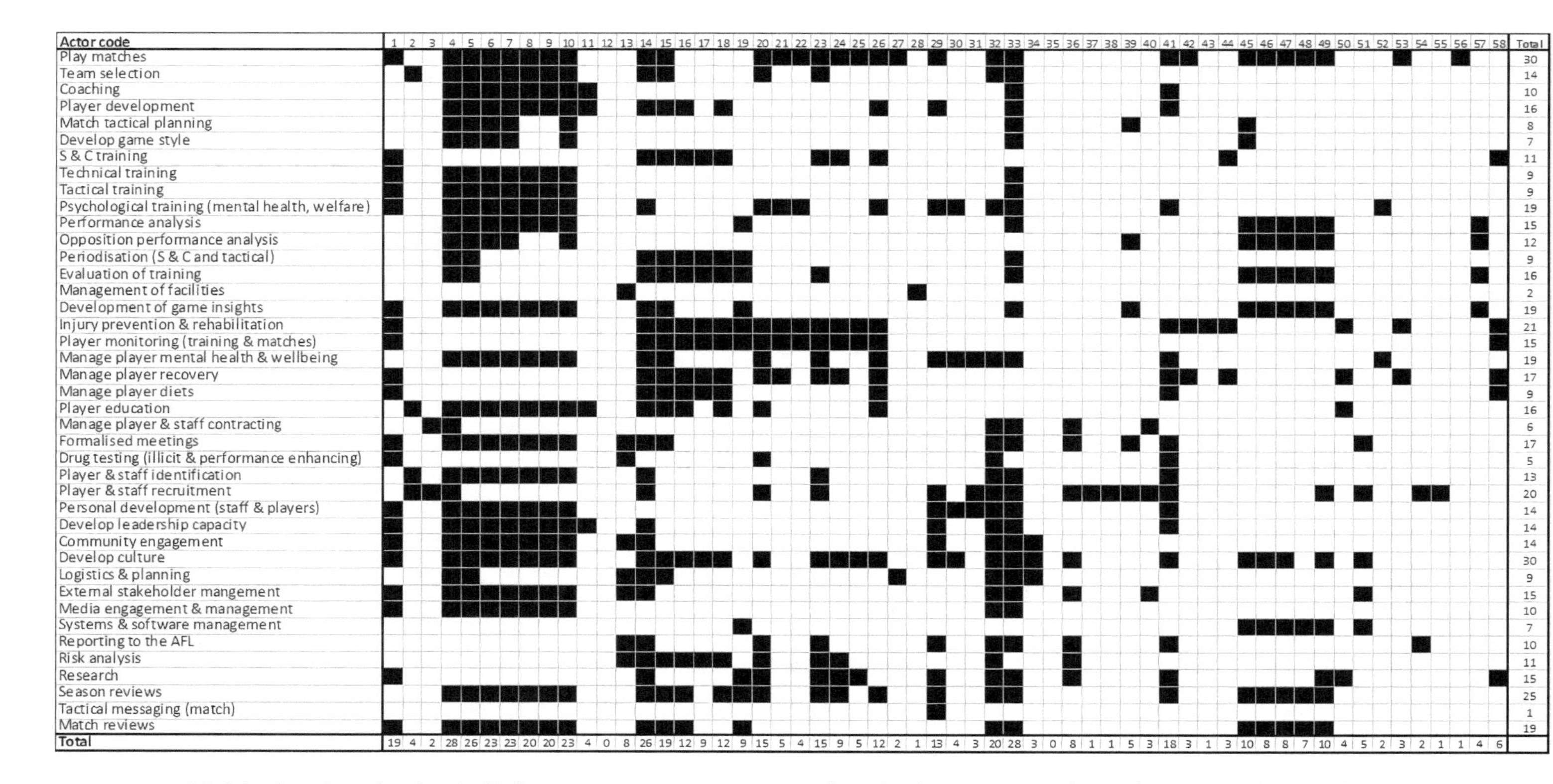

FIGURE 4.5 SOCA showing the football department actors associated with the purpose-related functions identified in the abstraction hierarchy.

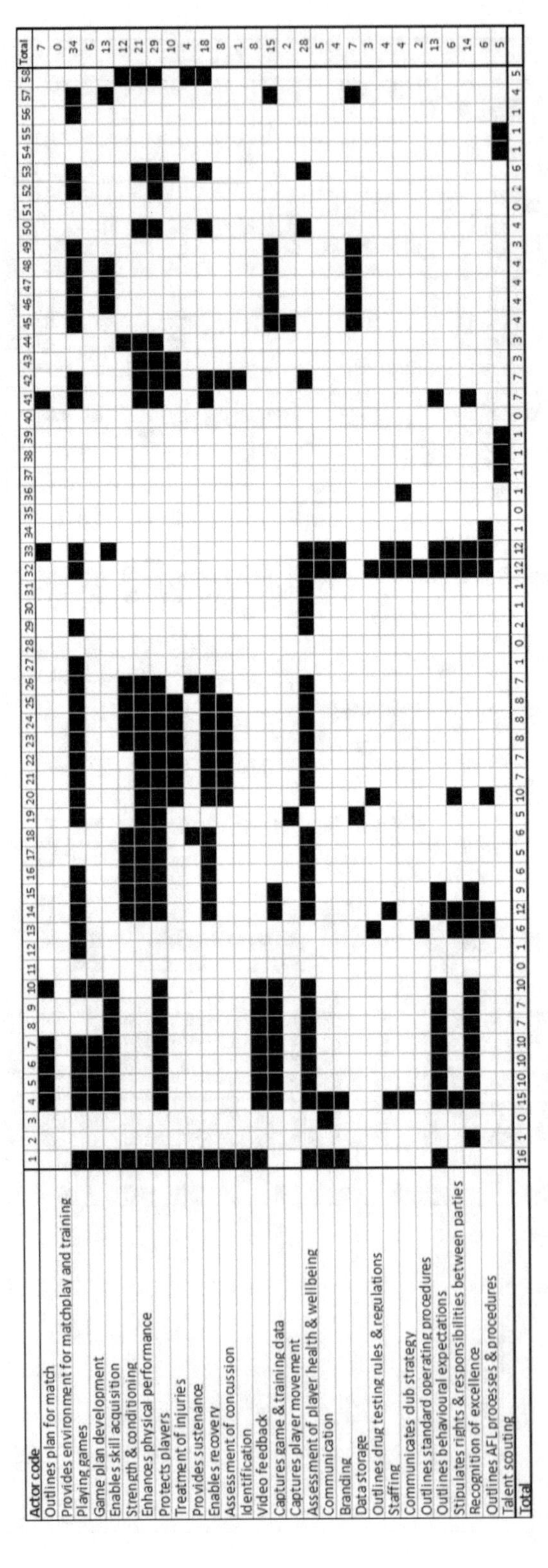

FIGURE 4.6 SOCA showing the football department actors associated with the object-related processes identified in the abstraction hierarchy.

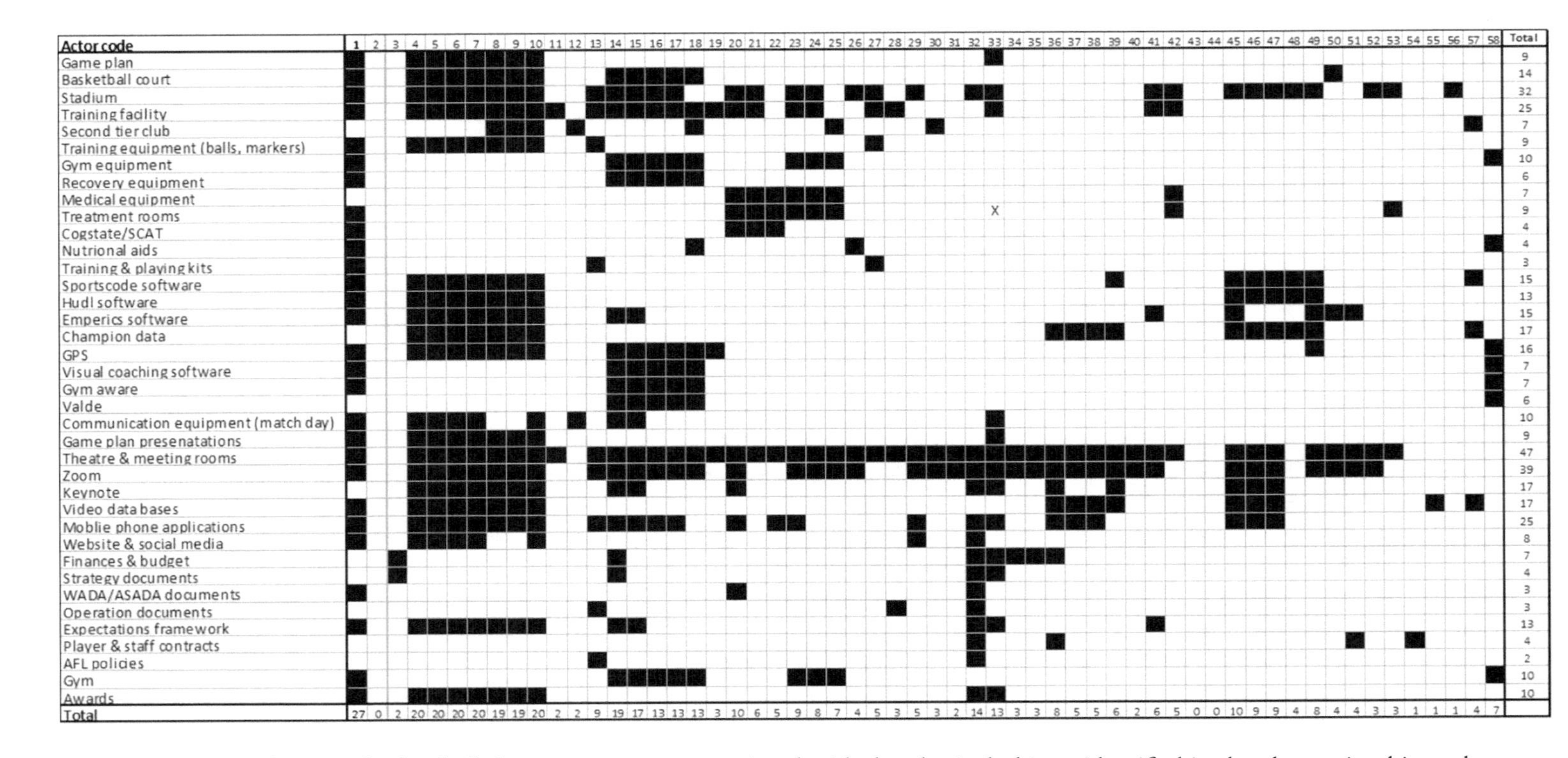

FIGURE 4.7 SOCA showing the football department actors associated with the physical objects identified in the abstraction hierarchy.

organisational cohesion, increase system resilience, reduce overlap, and reduce operational costs.

Recommended Reading

McLean, S., Solomon, C., Gorman, A., & Salmon, P. M. (2017). What's in a game? A systems approach to enhancing performance analysis in football. *Plos One*, 1–15.

McLean, S., Rath, D., Lethlean, S., Hornsby, M., Gallagher, J., Anderson, D., & Salmon, P. M. (2021). With crisis comes opportunity: Redesigning elite sports clubs for life after a global pandemic. *Frontiers in Psychology*, 11, 588959.

Jenkins, D. P., Stanton, N. A., Salmon, P. M., & Walker, G. H. (2009). *Cognitive work analysis: Coping with complexity*. Aldershot: Ashgate.

Stanton, N. A., Salmon, P. M., Walker, G. H., & Jenkins, D. P. (2017). *Cognitive work analysis: Applications, extensions and future*. Boca Raton, FL: CRC Press.

Vicente, K. J. (1999). *Cognitive work analysis: Toward safe, productive, and healthy computer-based work*. Mahwah, NJ: Lawrence Erlbaum Associates.

References

Ashoori, M., & Burns, C. (2013). Team cognitive work analysis: Structure and control tasks. *Journal of Cognitive Engineering and Decision Making*, 7, 123–140.

Bisantz, A. M., & Burns, C. M. (2008). *Applications of cognitive work analysis*. Boca Raton, FL: CRC Press, Taylor & Francis Group.

Berber, E., McLean, S., Beanland, V., Read, G. J., & Salmon, P. M. (2020). Defining the attributes for specific playing positions in football match-play: A complex systems approach. *Journal of Sports Sciences*, 38(11–12), 1248–1258.

Berber, E., Read, G. J., & Simcock, G. (2022). Sharing lessons learnt: Reflections on a novel approach to developing the contextual activity template. *Human Factors and Ergonomics in the Manufacturing & Service Industries*, 32(1), 50–65.

Burns, C. M., Bisantz, A. M., & Roth, E. M. (2004). Lessons from a comparison of work domain models: Representational choices and their implications. *Human Factors*, 46, 711–727.

Cornelissen, M., Salmon, P. M., McClure, R., & Stanton, N. A. (2013). Using cognitive work analysis and the strategies analysis diagram to understand variability in road user behaviour at intersections. *Ergonomics*, 56, 764–80.

Cornelissen, M., Salmon, P. M. & Young, K. L. (2013). Same but different? Understanding road user behaviour at intersections using cognitive work analysis. *Theoretical Issues in Ergonomics Science*, 14(6), 592–615.

Hulme, A., McLean, S., Read, G., Dallat, C., Bedford, A., & Salmon, P. M. (2019). Sports organisations as complex systems: Using cognitive work analysis to identify the factors influencing performance in an elite Netball organisation. *Frontiers Sports Acting Living - Sports Management and Marketing*, 1, 1–56.

Jenkins, D. P., Stanton, N. A., Salmon, P. M., Walker, G. H., Young, M. S., Whitworth, I.,... & Hone, G. (2007). The development of a cognitive work analysis tool. In Engineering Psychology and Cognitive Ergonomics: 7th International Conference, EPCE 2007, Held as Part of HCI International 2007, Beijing, China, July 22–27, 2007. Proceedings 7 (pp. 504–511). Springer Berlin Heidelberg.

Jenkins, D. P., Stanton, N. A., Salmon, P. M., & Walker, G. H. (2009). *Cognitive work analysis: Coping with complexity*. Surrey: Ashgate.

Jones, G., Gittins, M., & Hardy, L. (2009). Creating an environment where high performance is inevitable and sustainable: The high performance environment model. *Annual Review of High Performance Coaching Consulting*, 1(13), 139–150.

Klein, G., & Armstrong, A. A. (2005). Critical decision method. In N. Stanton, A. Hedge, K. Brookhuis, E. Salas & H. Hendrick (Eds.), *Handbook of human factors and ergonomics methods* (pp. 347–356). Boca Raton, FL: CRC Press.

Klein, G. A., Calderwood, R., & MacGregor, D. (1989). Critical decision method for eliciting knowledge. *IEEE Transactions on Systems, Man and Cybernetics*, 19(3), 462–472.

McLean, S., Solomon, C., Gorman, A., & Salmon, P. M. (2017). What's in a game? A systems approach to enhancing performance analysis in football. *Plos One*, 12(2), e0172565.

Mclean, S., Hulme, A., Mooney, M., Read, G. J., Bedford, A., & Salmon, P. M. (2019). A systems approach to performance analysis in women's netball: Using work domain analysis to model elite netball performance. *Frontiers in Psychology*, 10, 201.

McLean, S., Rath, D., Lethlean, S., Hornsby, M., Gallagher, J., Anderson, D., & Salmon, P. M. (2021). With crisis comes opportunity: Redesigning elite sports clubs for life after a global pandemic. *Frontiers in Psychology*, 11, 588959.

Naikar, N., Pearce, B., Drumm, D., & Sanderson, P. M. (2003). Designing teams for first-of-a-kind, complex systems using the initial phases of cognitive work analysis: Case study. *Human Factors*, 45, 202–217.

Naikar, N., Moylan, A., & Pearce, B. (2006). Analysing activity in complex systems with cognitive work analysis: Concepts, guidelines and case study for control task analysis. *Theoretical Issues in Ergonomics Science*, 7(4), 371–394.

Naikar, N. (2013). *Work Domain Analysis: Concepts, Guidelines and Cases*. Boca Raton, FL: Taylor & Francis Group.

Rasmussen, J. (1983). Skills, rules and knowledge – signals, signs and symbols, and other distinctions in human performance models. *IEEE Transactions on Systems, Man and Cybernetics*, 13, 257–266.

Rasmussen, J. (1987). *Information processing and human-machine interaction. An approach to cognitive engineering*. New York: North-Holland.

Rasmussen, J., Pejtersen, A. M., & Goodstein, L. P. (1994). *Cognitive systems engineering*. New York: Wiley-Interscience.

Read, G. J. M., Salmon, P. M., Goode, N., & Lenné, M. G. (2018). A sociotechnical design toolkit for bridging the gap between systems-based analysis and system design. *Human Factors in the Manufacturing and Services Industries*, 28(6), 327–341.

Salmon, P. M., Stevens, N. J., McLean, S., Hulme, A., & Read, G. J. M. (2021). Human Factors and Ergonomics and the management of existential threats: A work domain analysis of a COVID-19 return from lockdown restrictions system. *Human Factors and Ergonomics in the Manufacturing and Service Industries*, 31(4), 412–424.

Salmon, P. M., & McLean, S. (2020). Complexity in the beautiful game: Implications for football research and practice. *Science and Medicine in Football*, 4(2), 162–167.

Salmon, P. M., McLean, S., Carden, T., King, B. J., Thompson, J., Baber, C., ... & Read, G. J. (2024). When tomorrow comes: A prospective risk assessment of a future artificial general intelligence-based uncrewed combat aerial vehicle system. *Applied Ergonomics*, 117, 104245.

Stanton, N. A., Salmon, P. M., Walker, G. H., & Jenkins, D. P. (2017). *Cognitive work analysis: Applications, extensions and future*. Boca Raton, FL: CRC Press.

Vicente, K. J. (1999). *Cognitive work analysis: Toward safe, productive, and healthy computer-based work*. Mahwah, NJ: Lawrence Erlbaum Associates.

5

THE EVENT ANALYSIS OF SYSTEMIC TEAMWORK (EAST) AND EAST BROKEN LINKS (EAST-BL) APPROACH

Background

Teamwork in sport is often the deciding factor between successful and unsuccessful performance, be it on the pitch, the club rooms, the boardroom, or across all three. Given the increasing adoption of advanced technologies in sport, the nature of teams and teamwork is changing, with human-AI teams becoming the norm. Methods to help understand and optimise teamwork in sport are therefore critical. The Event Analysis of Systemic Teamwork (EAST) (Stanton et al., 2013; 2018) provides an integrated suite of methods for analysing the behaviour of teams, organisations, and sociotechnical systems (Salmon & Plant, 2022).

The EAST framework provides methods to describe, analyse and integrate three network-based representations of activity: task, social, and information networks (see Stanton et al., 2018). Task networks are used to model the interrelated tasks that are undertaken within a system. Social networks are used to describe the human and non-human actors performing these tasks and the communications and interactions that take place between them during task performance. Information networks describe the information that these actors use during task performance and how this information is distributed across tasks and actors. The three networks are subsequently analysed using a range of network analysis metrics (see Chapter 7) to examine network structure and the connectedness of different nodes within the networks. In a more recent iteration of the framework, the task, social, and information networks can be integrated to form composite networks showing the relationships between tasks, actors, and information (Stanton et al., 2018).

The EAST framework is particularly useful when attempting to understand situation awareness (SA) and how it is distributed throughout a system – a concept known as distributed situation awareness (DSA) (Salmon & Plant, 2022; Stanton et al., 2006; 2017). According to Stanton et al. (2006), DSA is an emergent property that is held by the overall system and is built through interactions between actors, both human and non-human (e.g., technology, tools, documents, displays). Stanton et al. argued that the SA required

DOI: 10.4324/9781003259473-7

for successful performance is not held by any one actor alone but instead resides in the interactions occurring across the sociotechnical system. A system's awareness therefore comprises a network of information upon which different components of the system have distinct views and ownership. Therefore when SA decrements result in sub-standard sports performance it is not because individual athletes or coaches lose SA; rather, it is a result of the sport system losing SA.

The EAST framework is particularly useful for understanding DSA across team members, either in athlete teams, match official teams, or coach and athlete teams (Hulme et al., 2021; Macquet & Stanton, 2014; Neville, 2020; Neville et al., 2016; Salmon et al., 2017). The implications for understanding DSA in sport are profound, suggesting for example that in a soccer match instead of having shared SA, each player will have their own unique understanding of the situation, as will the opposition players, match officials, coaches, fans, and technologies such as the video assistant referee (VAR). Each actor will also contribute different information to the systems DSA, with coaches communicating their intentions verbally, players communicating information both verbally and non-verbally, and the match officials through artefacts such as whistles, flags, and spray foam. Optimal performance of the soccer match 'system' is based on these different views connecting together, as well as 'transactions' in SA where the appropriate information is given to the right actor at the right time (Salmon et al., 2018). As such, each actor in the system needs to have compatible awareness but not necessarily the same (or shared) awareness as other actors. DSA has become an increasingly important concept in fields where distributed teams and complex systems are common (e.g., sport), and it has the potential to improve safety, efficiency, and overall performance. Further, as technology continues to advance, DSA is likely to become even more prevalent as teams will comprise non-human actors (e.g., technology) (Banks et al., 2018). The rapid pace of development and implementation of advanced technology is such that it will become an important area of future research and practice in sport.

The EAST-Broken Links (EAST-BL) (Stanton & Harvey, 2017) method is an extension to the EAST systems analysis framework that enables it to be used for prospective risk assessment purposes. Applying EAST-BL involves breaking the links in EAST task and social networks to identify the risks associated with failures in communication and information transfer. According to Stanton and Harvey (2017), broken links represent "failures in communication and information transfer between nodes in the networks and these failures can be used to make predictions about the possible risks within the sociotechnical system" (Stanton & Harvey, 2017). Initial testing of the approach in elite women's cycling demonstrate its utility for identifying risks to sports team performance (Hulme et al., 2021).

This chapter comprises three sections: (1) guidance on EAST applications, (2) guidance on the EAST-BL prospective risk assessment approach, and (3) a case study applying EAST and EAST-BL approach to understand DSA regarding player readiness within a football club performance department.

Applications in sport

Though first developed to support the analysis of defence and civilian command and control systems (Stanton et al., 2013), the EAST framework has since been applied in various sports contexts. These include the tasks, interactions, and information required during women's elite cycling road races (Salmon et al., 2017), to examine differences in coach and athlete SA (Macquet & Stanton, 2014), and to investigate DSA in AFL umpires (Neville, 2020; Neville et al., 2016). Applications of EAST and EAST-BL have potential to contribute to a range of contexts in sport, including human teaming, e.g., at a team, club, and organisational levels, but also to better understand DSA in human and non-human teams that are emerging within sport, e.g., AI injury prediction tools, and officiating with VAR.

Procedure and advice

A flowchart depicting a step-by-step process of the EAST procedure is presented in Figure 5.1.

Step 1: Define analysis aims

Firstly, it is important to establish the analysis aims and boundary to ensure the selection of suitable scenarios and collection of pertinent data. Furthermore, as not all elements of the EAST framework may be necessary, defining these aims from the outset is essential for the appropriate application of EAST elements (Salmon et al., 2022). For example, it is possible to use the information network as a standalone method to model and assess DSA (e.g., Salmon et al., 2016). Setting an appropriate analysis boundary is especially important, as this prevents data collection and analysis activities from going beyond what is required. Within sport, depending on the analysis aims the analysis boundary could be set at team, club, league, national, international, or entire sports system level.

Step 2: Define the task/scenario/system under analysis

In Step 2, it is important to clearly define the task(s), scenario(s), or system you wish to analyse. EAST is a flexible framework which can be used to describe and analyse the behaviour of individuals (e.g., Salmon et al., 2014), teams (e.g., Roberts et al., 2017), organisations (e.g., Walker et al., 2010) and entire sociotechnical systems (e.g., Stanton et al., 2019). It isessential to clearly define the focus and limits of the analysis, ensuring all analysts are aware of what is and isn't included within scope. The scope should be carefully considered to avoid being overly narrow, which could limit the systems thinking perspective, but also to prevent it from being so broad that it becomes unmanageable. The EAST method offers flexibility, allowing for the scope and boundaries of the analysis to be revisited and potentially adjusted if new relevant data and/or topics are identified during the analysis process.

Step 3: Data collection

Step 3 involves collecting data about the system and the activities of interest. Data collection for EAST typically involves a combination of documentation review (e.g., standard operating procedures), direct observations of the tasks/scenarios of interest, gathering recordings of communications between actors, e.g., AFL umpires (Neville et al., 2016), eliciting concurrent verbal protocols from the actors performing tasks, or conducting post

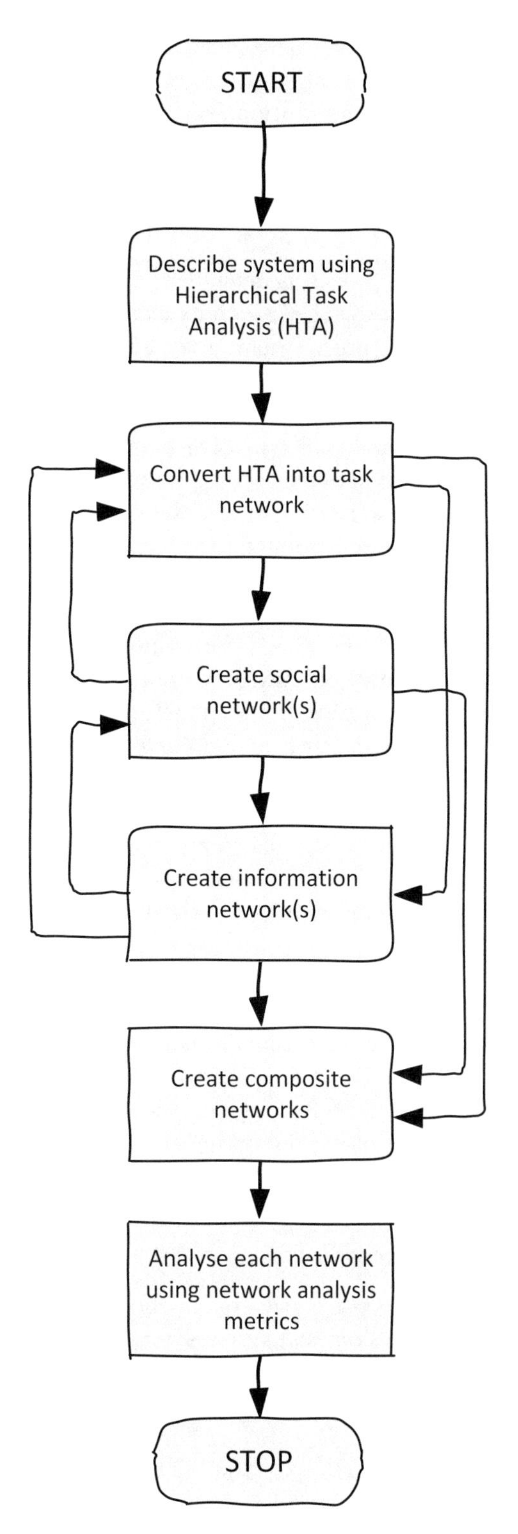

FIGURE 5.1 EAST procedure flowchart.

hoc structured or semi-structured interviews, e.g., Cognitive Decision Method (CDM) (O'Hare, 2000). An example of a post hoc data collection method in sport is McLean and colleagues (2018) Intra-Team Communication Tool, which can be used post-hoc to gain subjective data on the amount of communication soccer players received from other team members, as well as how beneficial that communication was to their match performance. For more detailed information on data collection methods for EAST the reader is referred to Stanton et al. (2013).

The chosen data collection approach is dependent upon the aims and boundaries of the analysis as well as project constraints, such as available resources (time, analysts), and access to data sources, e.g., sports system actors, or observational data. Typically, multiple analysts observe the activities under analysis, each focusing on either individual actors, or for larger analyses a sub-set of actors. For example, an EAST analysis of player development in a sports club might only involve a sub-set of actors and not the entire performance department. All activities should be recorded, and an activity timeline should be created, including a description of the activity undertaken, the actors involved, any communications made between actors and the tools, equipment and technology involved. Additional notes should be made where required, including the purpose of the activities observed, any documents or instructions used to support each activity, activity outcomes, any errors made, and any other information that the actors involved feel is relevant. It is normal practice, where permitted, to make video recordings of the activities and audio recordings of communications between actors.

Conducting CDM interviews with actors undertaking the activities under analysis is particularly useful (Salmon et al., 2022). This involves breaking down the activity or scenario into key phases and then interviewing each actor using a set of pre-defined CDM probes. Pre-defined CDM probes include goal specification; assessment; cue identification; expectancy; options; influencing factors; situation awareness; situation assessment; experience; mental models; decision making; guidance; and basis of choice (see O'Hare et al., 2000).

Step 4: Transcribe data

Once collected, the data should be transcribed (if necessary). When transcribing the data it is recommended that an event transcript be constructed. This should describe the activity over a timeline, including descriptions of tasks, the actors involved, any communications made, and the technology used (Salmon et al., 2022). The event transcript data should be reviewed by SMEs to ensure the validity of the data.

Step 5: Construct HTA for task or scenario under analysis

During Step 5, a HTA for the task/scenario/system should be developed. HTA is a task analysis method that is used to decompose systems, tasks, or behaviours into a hierarchy of goals, sub-ordinate goals, operations, and plans (see Chapter 3). The completed HTA should be reviewed by appropriate SMEs following completion. The HTA component is particularly important as it is used to identify activity phases for which task, social, and information networks will be constructed. The HTA also acts as a useful input to the task and social networks as it specifies what tasks are undertaken and what actors are interacting with one another when undertaking them.

Step 6: Define activity or scenario phases

In Step 6, the HTA should be reviewed to determine whether the analysis should focus on an overall activity, or a set of discrete activity phases. For example, when examining a cycling road race the EAST analysis could focus on the overall task of 'finishing the race with the protected rider in a winning position', or instead on discrete cycling race activities or phases, such as the' warm up', 'race tactics', or 'mounting an attack' (Hulme et al., 2021). Alternatively, the analysis could focus on phases such as 1. Race planning and tactics, 2. Warm up, 3. Provide mechanical or nutritional support, 4. Race start, 5. Establish/maintain appropriate position in peloton, 6. Form break or attack, 7. Sprint finish, and 8. Control convoy. The specific focus is dependent on the analysis aims and boundaries and the quality of the data collected. If the analysis is to be an analysis of the activity in question, then an overall task, social and information network should be constructed during Steps 7–9. If there are a set of activity phases, then task, social, and information networks should be constructed for each activity phase.

Steps 7-9: Construct EAST networks

Steps 7–9 provide guidance on developing task, social, and information networks. A set of guidelines for constructing the networks are presented in Table 5.1. It is recommended that each network is created first in Microsoft Excel in the form of a matrix and then built in a network analysis software program (see Chapter 7 – Network analysis, for additional guidance on building network matrices and using network analysis programs). The networks can also be drawn in software packages such as Microsoft Visio. In Table 5.1, examples are used from the case study presented later in the chapter on DSA within a football club's performance department regarding player readiness.

TABLE 5.1 Analysis rules regarding the relationships between nodes within EAST networks (adapted from Salmon et al., 2022)

Network	*Nodes*	*Example nodes*	*Relationships*	*Example relationships*
Task network	Represent high-level tasks that are undertaken during the activity under analysis. The high-level tasks are typically extracted from the sub-ordinate goals level of the HTA.	– S & C training – Load management – Player monitoring	Represent instances where the conduct of one task (i.e., task network node) influences, is undertaken in combination with, or is dependent on, another task.	Load management influences the decisions around S & C training.

(*Continued*)

TABLE 5.1 (Continued)

Network	*Nodes*	*Example nodes*	*Relationships*	*Example relationships*
Social network	Represent human, technological, or organisational actors who undertake one or more of the tasks involved in the activity under analysis.	– Head coach – Sports scientist – Physiotherapist – Devices/sensors	Represent instances where actors within the social network interact with one another during the activity under analysis.	The Physiotherapist interacts with the head coach and/or the sport scientist on player injury status.
Information network	Represents pieces of information that are required by actors when undertaking the activity under analysis.	– Fatigue level – Recovery status – Injury status – Training schedule – Exercise intensity	Represent instances where information influences other information or is used in combination with other information during the activity under analysis.	Exercise intensity is linked to fatigue and recovery as this information is used to understand training adaptations.

Step 7: Construct task network

The first analysis step involves constructing a task network for the activity under analysis. Task networks (see Figure 5.2) are used to represent HTA outputs in the form of a network which shows key tasks and the relationships between them (Stanton et al., 2013; 2018). This enables analysts to understand the interactions and coupling between tasks. Within the case study example of a football club performance department (Figure 5.2), the circular nodes represent tasks and the arrows linking the tasks represent relationships between tasks. Stanton et al. (2018) describe four kinds of relationships between tasks that are included in the task network:

1 Tasks are undertaken sequentially, e.g., task three is undertaken after completion of task two;
2 Task are undertaken together, e.g., tasks one and two are undertaken together;
3 The outcomes of one task influence the conduct of another, e.g., the outcomes of task four influence how task five is undertaken; or
4 The conduct of one task is dependent on completion of the other, e.g., task five cannot be undertaken until task four is completed.

Task networks are typically constructed by taking the first layer of sub-goals from the HTA and identifying which of them are related with one another based on the relationship types described above. Once the initial draft task network is complete, it is useful to have SMEs review it. The task network should then be refined based on the SMEs feedback. Changes at this stage typically involve the addition and/or removal of relationships

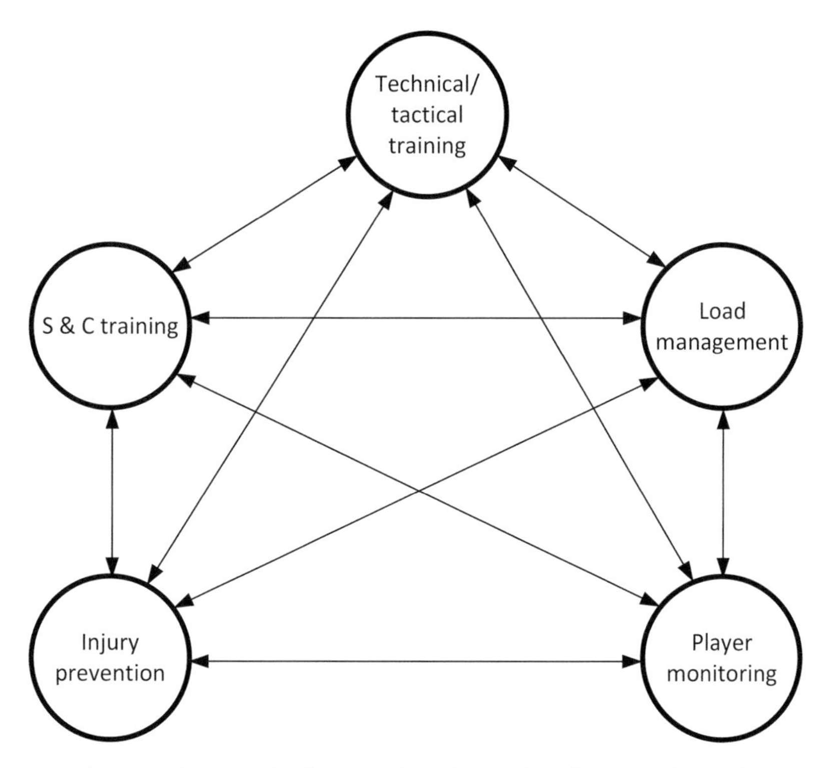

FIGURE 5.2 Task network example showing the relationships between five tasks required by a football performance department for player readiness.

between tasks or the modification of task descriptions. Like HTA, it is normal practice for the task network to go through various iterations before it is finalised.

We recommend that the development of a task network should be based on HTA, however, it is possible to develop a task network based directly on raw data collection activities, SME input and verification, and/or document review (Stanton, 2014). This is often a useful approach where project constraints mean there are not sufficient resources to support the development of a HTA. If the HTA step is not utilised, analysts should use SME review to ensure the validity of the task network.

Step 8: Construct social network

Construction of the social network involves identifying and depicting the relationships between actors in the social network. This is done by constructing a social network adjacency matrix that includes the frequency of interactions between actors in the system. For example, in a football club performance department the head coach and coaches interact daily around training design, etc., whereas the head coach and team psychologist may interact less frequently (Table 5.2). Typically, the direction (i.e., from actor A to actor B) and frequency of communications are included; however, the type and content of communications can also be included, as well as the communications media used (e.g., Walker et al., 2010). The example social network presented in Figure 5.3 represents the interactions between football club performance department actors – both human and non-human.

TABLE 5.2 Adjacency matrix of a social network in a football club performance department

	Head coach	*Coaches*	*Sports scientists*	*Psychologist*	*Match analyst*	*Devices/ sensors (e.g., GPS)*
Head coach	0	3	3	2	3	0
Coaches	3	0	3	2	3	0
Sports scientists	3	3	0	2	2	3
Psychologist	1	0	1	0	0	0
Match analyst	3	3	2	0	0	2
Devices/sensors (e.g., GPS)	0	0	3	0	2	0

Note: To represent strength of connection the values in matrix represent the frequency of interactions, 3 = daily, 2 = weekly, 1 = monthly.

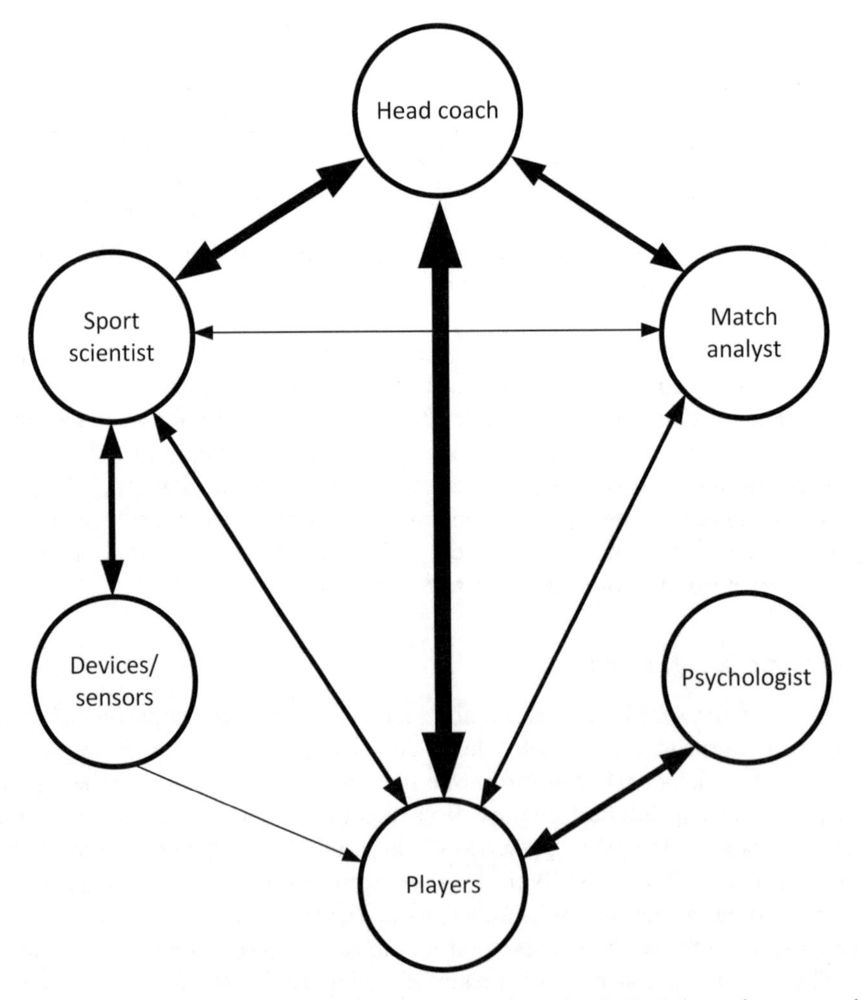

FIGURE 5.3 Social network diagram for a selection of six football club performance department actors regarding their interactions. Arrow thickness depicts more frequent interactions.

Step 9: Construct information network

The information network, or SA network, is the final network to be constructed within EAST analyses. The information network Comprises the information used by the actors in the social network when conducting the tasks described within the task network (Stanton et al., 2018). Construction of the information networks can be performed manually or using software tools such as Leximancer based on transcribed data provided by participants performing the tasks under analysis. In the absence of transcribed data from communication logs or concurrent verbal protocols information networks can also be developed based on procedures, or reports (e.g., Salmon et al., 2016).

Information networks depict the information or concepts underlying SA and the relationships between them (Stanton et al., 2018). For the football club performance department case study, exercise intensity contributes to fatigue and recovery status, and influences exercise duration. This is shown below in the form of an information network in Figure 5.4.

Manual construction of the networks involves performing a content analysis on the data to identify concepts and the relationships between them. It is often useful to have a second analyst also construct an information network (or at least a portion of it) from the same data for the purposes of reliability testing.

Step 10: Construct composite networks

Composite networks are used to explore the relationships between tasks, actors, and information (Stanton, 2014). Composite networks are constructed by combining the task, social and information networks in different ways. For example, a task-social composite

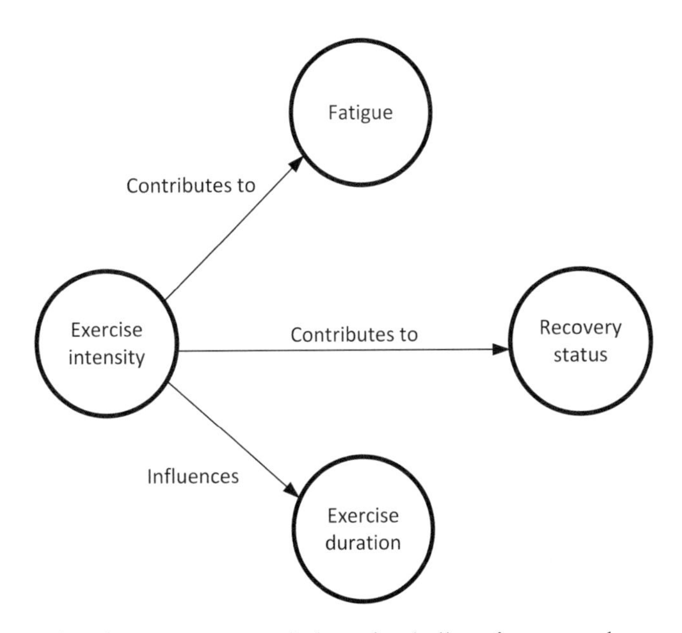

FIGURE 5.4 Example information network for a football performance department for assessing player readiness.

network can be constructed by combining the task and social network to show which tasks are undertaken by which actors. A task-information network can be constructed by combining the task and information networks to identify which information is required to undertake each task. A social-information network can be constructed by combining the social and information networks to identify which information is required to be transferred between actors for specific tasks. Various composite networks can be constructed (Table 5.3):

Figure 5.5 demonstrates a task-information and a social-information composite networks whereby information from the information network is transferred between tasks and actors. Alternatively, the composite networks can be developed in tabular format (Tables 5.4 and 5.5).

TABLE 5.3 Composite network types and structure

Composite network	*Structure*
Task-actor network	Combined task and social network
Social-information network	Combined social and information network
A task-information network	Combined task and information network
Fully composite network showing information by actors and related to tasks	Combined task, social, and information network (see Salmon et al., 2017)

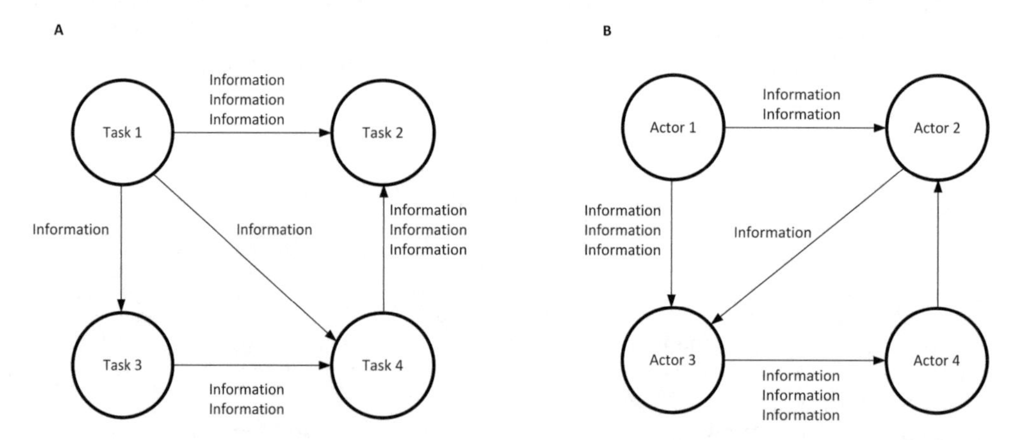

FIGURE 5.5 Task-information network (Panel A) and social-information network (Panel B).

TABLE 5.4 Tabularised task-information composite networks

From (Task)	*To (Task)*	*Information required*
Task 1	Task 2	Information Information Information
Task 3	Task 4	Information Information

TABLE 5.5 Tabularised social-information composite networks

From (Actor)	*To (Actor)*	*Information required*
Actor 1	Actor 2	Information Information
Actor 3	Actor 4	Information Information Information

Once the task, social, information, and composite networks are complete, it is pertinent to validate the outputs using appropriate SMEs, and where possible recordings of the scenario under analysis. The composite networks are necessary for performing the EAST-BL phase.

Step 11: Analyse networks using network analysis metrics

The next step involves analysing the networks using appropriate network analysis metrics. Various metrics exist, with the specific metrics used being dependent on the analysis aims. For example, nodal metrics can be used to look at individual behaviour and status within the network, whereas overall network metrics can be used to look at the structure of the entire network (Mclean et al., 2023). Table 5.6 presents a summary of common network metrics and their definitions and applications in sport. It should be noted that Table 5.6 is not an exhaustive list of network analysis metrics, but a brief guide to a selection of commonly applied metrics in sport. One method to identify key nodes within networks is to identify those that are one standard deviation above the mean value for each network metric (Baber et al., 2006; McLean et al., 2023). This is often useful to identify key tasks, actors, or pieces of information.

Training and application times

Due to the number of different methods involved, the training time associated with the EAST framework is high and involves understanding how to construct and analyse the three initial networks and corresponding composite networks. Similarly, application time is typically high, although this is dependent upon the task under analysis and the analysis aims and boundary (Salmon et al., 2022).

Reliability and validity

The reliability and validity of the EAST methods are difficult to assess and at the time of writing there have been no formal EAST reliability and validity studies (Salmon et al., 2022).

Tools needed

There is no dedicated EAST software tool, however, there are various tools which can be used to help build and analyse the networks. For data collection, video and audio recording devices are used to record the activities under analysis. Network matrices are typically

TABLE 5.6 Network analysis metrics examples

Network metric	*Definition*
Network density	Network density calculates the proportion of actual connections (edges) in a network compared to the total possible connections.
Out-degree centrality	Out-degree centrality is the count of outgoing connections that a node initiates towards other nodes in a network. It helps measure the proactivity, influence, or connectivity tendencies of a specific node within the network.
In-degree centrality	In-degree centrality is the count of incoming connections that a node receives from other nodes in a network. It helps measure the popularity, influence, or dependency on a specific node within the network.
Betweenness centrality	Betweenness centrality quantifies the extent to which a node lies on the shortest paths between other nodes. It helps identify key actors who play a crucial role in connecting other nodes, e.g., an agent who acts as abridge bridge.
Closeness centrality	Closeness centrality measures how quickly a node can access other nodes in the network. It helps identify actors who are in close proximity to other actors and can connect efficiently.
Eigenvector centrality	Eigenvector centrality measures a node's importance based on the importance of its connections. It helps identify key actors who have strong connections to other influential actors, indicating their influence or centrality within the network.
Clustering coefficient	The clustering coefficient measures the extent to which nodes in a network tend to cluster together. It helps identify cohesive units within a team or actors who tend to interact within their group.

created in Microsoft Excel, with the corresponding network diagrams drawn using tools such as Microsoft Visio, and PowerPoint. For developing network visualisations, it is recommended that network analysis programs and/or computational programs be used to conduct network analysis through online tools, such as Gephi, R and igraph, SocNetV, NodeXL, among many others.

EAST-BL approach

The EAST-BL (Stanton & Harvey, 2017) method is an extension to the EAST systems analysis framework that enables it to be used for prospective risk assessment. As noted earlier, applying EAST-BL involves breaking the links in EAST task, social and information networks to identify failures in communication and information transfer and any risks associated with them. The practical guidance offered below is intended for instances where the EAST networks have been developed based on steps 1 – 11 above.

EAST-BL applications in sport

EAST-BL was originally developed and applied to identify the risks associated with a maritime missile attack training system (Stanton & Harvey, 2017); however, the approach is generic and can be applied in any domain. For example, EAST-BL has recently been applied to identify the risks to performance in elite women's road cycling (Hulme et al., 2021).

Practical guidance for EAST-BL

If applying EAST-BL, the analyst will be required to undertake additional steps after completing an initial EAST analysis. A flowchart depicting a step-by-step process of the EAST-BL procedure is presented in Figure 5.6.

Step 12: Break links in task-information networks

Steps 12 and 13 describe the process of breaking the links in the task-information and social-information networks respectively. Once the three EAST networks are finalised, the task-information network is subjected to the EAST-BL process. This involves systematically breaking each of the relationships between the tasks in the task network and identifying what risks emerge when the relevant information from the information network is not transferred between tasks. For example, in the football club performance department case study task network, the two tasks 'player monitoring' and 'load management' would be related as monitoring of the player is required to monitor load on

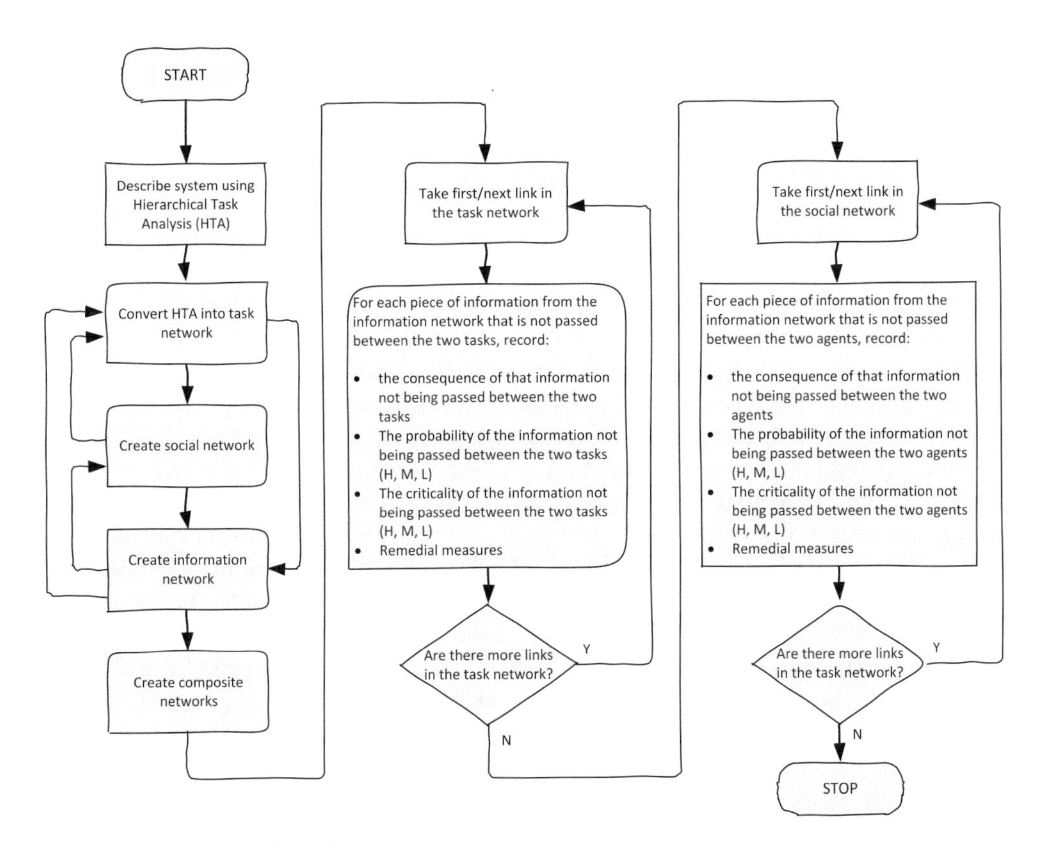

FIGURE 5.6 EAST-BL flow chart.

TABLE 5.7 Task-information network EAST-BL template

From (Task)	*To (Task)*	*Information not transferred*	*Outcome*	*Remedial measure*
Player monitoring	Load management	Exercise intensity	Coaches and support personnel will not be able to monitor the load imposed on the players, resulting in overload and injury, or underload and fitness losses.	Use of devices and sensors that capture exercise intensity.

the player. Applying EAST-BL, the analyst would break the relationship and then determine what the outcome would be if relevant information from the information was not transferred between the two tasks. In this case, information regarding 'exercise intensity' would not be available when performing the task 'load management' as the information is not transferred from the 'player monitoring' task. A consequence of this may be that the coaching and support personnel would not know the load imposed on the player. It is important to note that, when applying EAST-BL, analysts do not need to justify breaking any of the relationships or provide reasons why the relationship is broken. Rather, the process involves systematically breaking every relationship in the task and social network and identifying the associated risks. A task-information network EAST-BL output table for player load example is presented in Table 5.7.

Step 13: Break links in social-information network

Once all task-information network breaks have been assessed, the social-information network is subjected to the EAST-BL process. This involves systematically breaking each of the relationships between the actors in the social network and identifying what risks emerge when the relevant information from the information network is not transferred between actors (Salmon et al., 2022). It is important to note here that actors can be both human and non-human. For example, in the football club performance department case study, the two actors 'sports scientist' and 'head coach' and would be related, as would 'devices/sensors' and 'sport scientist's' as they share information on numerous performance aspects within a football club. Applying EAST-BL, the analyst would break the relationship between both actors and then determine what the outcome would be if relevant information from the information network was not transferred between them. For example, the information 'exercise intensity' would not be transferred from the sport scientist to the head coach. A consequence of this may be that the head coach will not know what intensity their planned training sessions are eliciting, which may result in fatigued players and subsequent injury and/or poor performance. The process involves systematically breaking every relationship in social-information network and identifying the associated risks. A social-information network EAST-BL output table for the information 'exercise intensity' example is presented in Table 5.8.

TABLE 5.8 Social network broken links template

From (Agent)	*To (Agent)*	*Information not transferred*	*Outcome*	*Remedial measure*
Sport scientist	Head coach	Exercise intensity	Coaches will not know what intensity their planned training sessions are eliciting, resulting in fatigued players and subsequent injury and poor performance.	Regular meetings between sports scientist and head coach to design training with regard to known exercise intensities the session elicit

Step 14: Identify remedial measures

The final stage in the EAST-BL process is to propose suitable controls for the identified risks. It is recommended that a group of analysts alongside SMEs be used to identify appropriate risk controls (Salmon et al., 2022). The group should work through the task-information and social-information network risks and discuss risk controls that can be used to either prevent the risk from occurring or mitigate the consequences of the risk. According to Stanton (2005), remedial measures proposed following risk assessment applications should be Considered, where possible, a range of categories including equipment; education and training; policy and procedures; organisational; and governance at a local, national, and international level.

Step 15: Review and refine analysis

Once the initial EAST-BL analysis is complete it is useful to have various SMEs review it. The analysis should then be refined based on SME feedback. This typically involves removing identified risks, identifying new risks, or adding or modifying risk management strategies.

Related methods

EAST-BL requires an initial EAST analysis comprising task, social, and information networks. EAST typically requires an initial HTA of the system under analysis, with the HTA subsequently used to support development of a task network for the system under analysis.

Approximate training and application times

EAST-BL is a relatively simple method that requires little training. In a recent reliability and validity study participants were provided with training that lasted around 2 hours and were able to achieve acceptable levels of validity (Hulme et al., 2021). Application time depends on the system under analysis; however, EAST-BL analyses typically require significant time as the analysis covers overall systems rather than merely sharp-end performance.

Reliability and validity

Hulme et al. (2021b) tested the criterion-referenced validity of EAST-BL, STAMP-STPA (Chapter 9) and Net-HARMS (Chapter 10). This involved training novice participants in one of the three methods and then asking them to use their respective methods to identify the risks across a task. Hulme et al. (2022) found moderate levels of validity for EAST-BL, suggesting that participants were able to achieve moderate levels of agreement with an expert 'gold standard' analysis.

Tools needed

EAST-BL can be conducted using pen and paper; however, analyses are normally undertaken using a Microsoft Excel spreadsheet (Salmon et al., 2022).

Advantages of EAST and EAST-BL

- As EAST analyses cover tasks, actors, and information (and the relationships between all three) it provides a comprehensive overview of the activities under analysis.
- EAST provides both a qualitative (network content) and quantitative (network analysis) analysis.
- EAST is highly flexible and can be applied to study team, club, organisational, and system activity in any sport.
- The use of network analysis metrics is a useful feature that enables the identification of key tasks, actors, and information and/or to compare activities in different contexts.
- The EAST framework is flexible and allows data collection and analysis methods to be chosen based on analysis requirements.
- The EAST networks enable analysis of various concepts, including DSA, distributed cognition, teamwork, risk, decision-making, workload, and human-AI teaming (Salmon et al., 2022).
- Software support is available for developing and analysing EAST networks.
- EAST-BL can be used proactively to identify risks across overall systems. This includes risks to performance in the sports context.
- EAST-BL analyses provide risk management strategies for each of the risks identified.
- EAST-BL has shown some reliability evidence (Hulme et al., 2022).

Disadvantages of EAST and EAST-BL

- EAST analyses can become large, complex, and unmanageable, especially when applied to overall systems.
- Though EAST and EAST-BL have been applied extensively across the safety critical domains, sport-specific analyses are only just emerging.
- Undertaking the full EAST and EAST-BL analyses can be highly time-consuming.
- EAST and EAST-BL can be difficult to apply for novel analysts.
- There is no reliability and validity evidence associated with EAST.
- The use of various data collection and analysis methods ensures that the time required for training and application can be higher than other systems analysis methods.
- A high level of access to the domain, activities, and relevant SMEs is often required.

Case study: Distributed situation awareness in a football club performance department

Distributed Situation Awareness is a concept that pertains to how individuals, teams, organisations, and sociotechnical systems understand, comprehend, and forecast system state (Salmon et al., 2019). In environments such as a football club's performance department, no single actor (human or non-human) holds all the relevant information required to assess and understand player readiness (to play and train). Player readiness is a holistic concept and requires a coordinated approach involving coaches, sports scientists, doctors, wellbeing specialists, and the athletes, among many others, for optimising performance, reducing injury risk, and ensuring long-term athlete development (Guthrie et al., 2023). The challenge lies in effectively sharing and integrating this information to form a comprehensive understanding between actors that can be leveraged for strategic decisions and performance improvements. This integration is crucial, considering the diverse roles and expertise within a football club's performance team (Mclean et al., 2021). By having a compatible understanding of the situation, individuals and teams can work together more effectively and make better informed decisions. The study of DSA in sport is in its infancy (e.g., Macquet & Stanton, 2014; Neville, 2020; Neville et al., 2016), however, given the increasing complexity of football performance, numerous stakeholders involved (including technology), time pressures within club settings, and more recently remote match officials, understanding DSA in football club performance teams is critical to support efforts to optimise performance. In this case study we demonstrate how the EAST framework can be used to understand and enhance DSA within a football club's performance department. We then apply EAST-BL as a risk assessment tool to identify risks associated with failures in communication and information transfer between performance department tasks and actors (Stanton & Harvey, 2017).

Previous peer-reviewed literature was drawn upon to depict the functioning and composition of a football performance department (McLean et al., 2021). Four network metrics were applied, Density, Out-degree centrality, In-degree centrality, and Closeness centrality. See Table 5.9 for network metric definitions.

EAST

The findings from the task network (Figure 5.7, Table 5.9) indicate that key tasks for player readiness include injury prevention, education, player monitoring, and health & wellbeing management.

The findings from the social network (Figure 5.8, Table 5.10) indicate that key actors for player readiness include the players themselves, sports scientists, and coaches.

The findings from the information network (Table 5.11, Figure 5.9) indicate that key pieces of information for player readiness include training and competition schedule, fatigue, recovery status, and tactics.

EAST-BL

An extract of the EAST-BL analysis is presented in Tables 5.12 and 5.13 to demonstrate the risks associated with player readiness when information is not transferred between tasks and actors. The identified risks then allow for remedial measures to be developed to mitigate the risks.

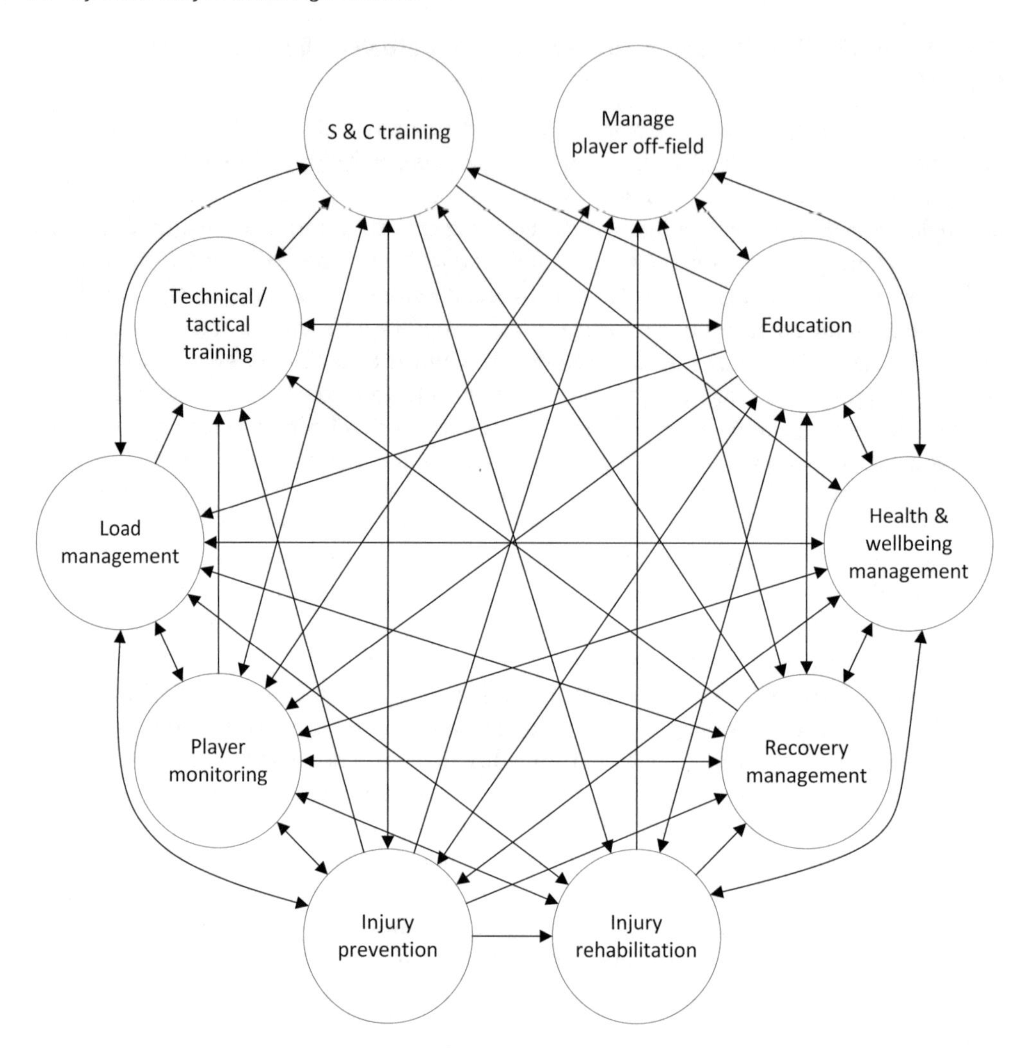

FIGURE 5.7 The task network comprised 10 tasks contributing to player readiness and were highly connected which was indicated through the network density (.72).

TABLE 5.9 Task network metrics. Shaded nodes represent those that are one standard deviation above the mean value for each metric and are thus deemed to be key nodes

Task	*Out-degree centrality*	*In-degree centrality*	*Closeness centrality*
S & C training	.66	.66	.75
Technical/tactical training	.22	.66	.56
Load management	.77	.77	.81
Player monitoring	.88	.88	.90
Injury prevention	1.0	.55	1.0

(*Continued*)

TABLE 5.9 (Continued)

Task	*Out-degree centrality*	*In-degree centrality*	*Closeness centrality*
Injury rehabilitation	.66	.66	.75
Recovery management	.77	.77	.81
Health & wellbeing management	.77	.88	.81
Education	1.0	.66	1.0
Manage player off-field	.44	.66	.64
Average	.71	.71	.80
Standard deviation	.24	.11	.14

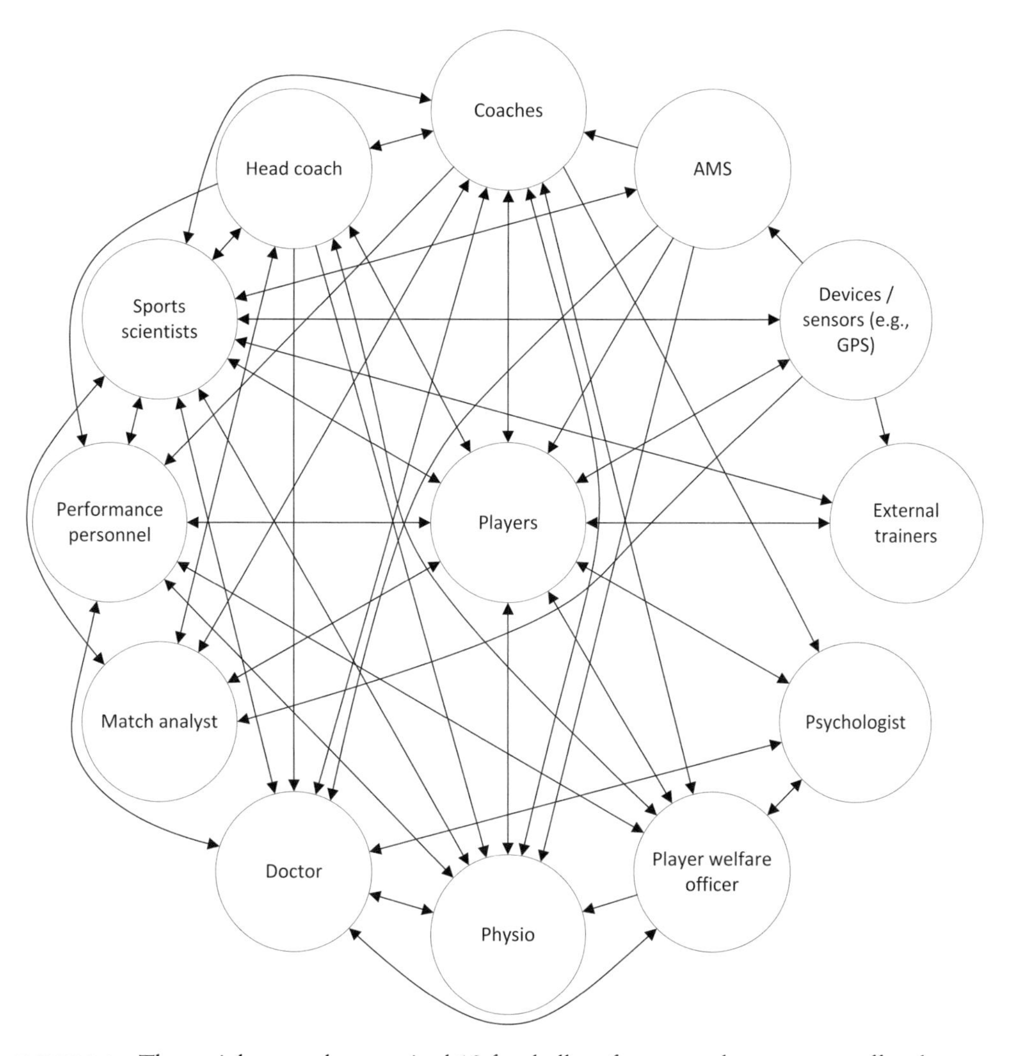

FIGURE 5.8 The social network comprised 13 football performance department staff and represents a moderately connected network indicated through the network density (.51).

TABLE 5.10 Social network metrics. Shaded nodes represent those that are one standard deviation above the mean

Actor	*Out-degree centrality*	*In-degree centrality*	*Closeness centrality*
Head coach	.66	.41	.75
Coaches	.75	.66	.80
Sport scientist	.83	.83	.85
Performance personnel	.41	.58	.63
Match analyst	.33	.41	.60
Doctor	.58	.75	.70
Physio	.41	.66	.63
Player welfare officer	.58	.50	.66
Psychologist	.25	.33	.54
External trainers	.16	.25	.54
Devices/sensors (e.g., GPS)	.41	.16	.63
Athlete management system	.41	.16	.63
Players	.96	1.0	.92
Average	.52	.52	.68
Standard deviation	.23	.26	.12

TABLE 5.11 Information network metrics. Shaded nodes represent those that are one standard deviation above the mean

Information	*Out-degree centrality*	*In-degree centrality*	*Closeness centrality*
Exercise duration	.45	.63	.00
Exercise intensity	.45	.63	.00
Exercise mode	.45	.63	.00
Exercise frequency	.45	.63	.00
Recovery status	.54	.72	.00
Fatigue	.54	.72	.00
Tactics	.18	.18	.44
Wellbeing	.36	.18	.00
Injury status	.09	.18	.00
previous injury status	.09	.18	.00

(*Continued*)

TABLE 5.11 (Continued)

Information	*Out-degree centrality*	*In-degree centrality*	*Closeness centrality*
Training schedule	.72	.18	.68
Competition schedule	.72	.18	.68
Average	.42	.42	.15
Standard deviation	.21	.25	.28

Note: Closeness centrality values of .00 indicate that the information is part of a segregated group (e.g., cluster) that is mostly separated from the main network structure.

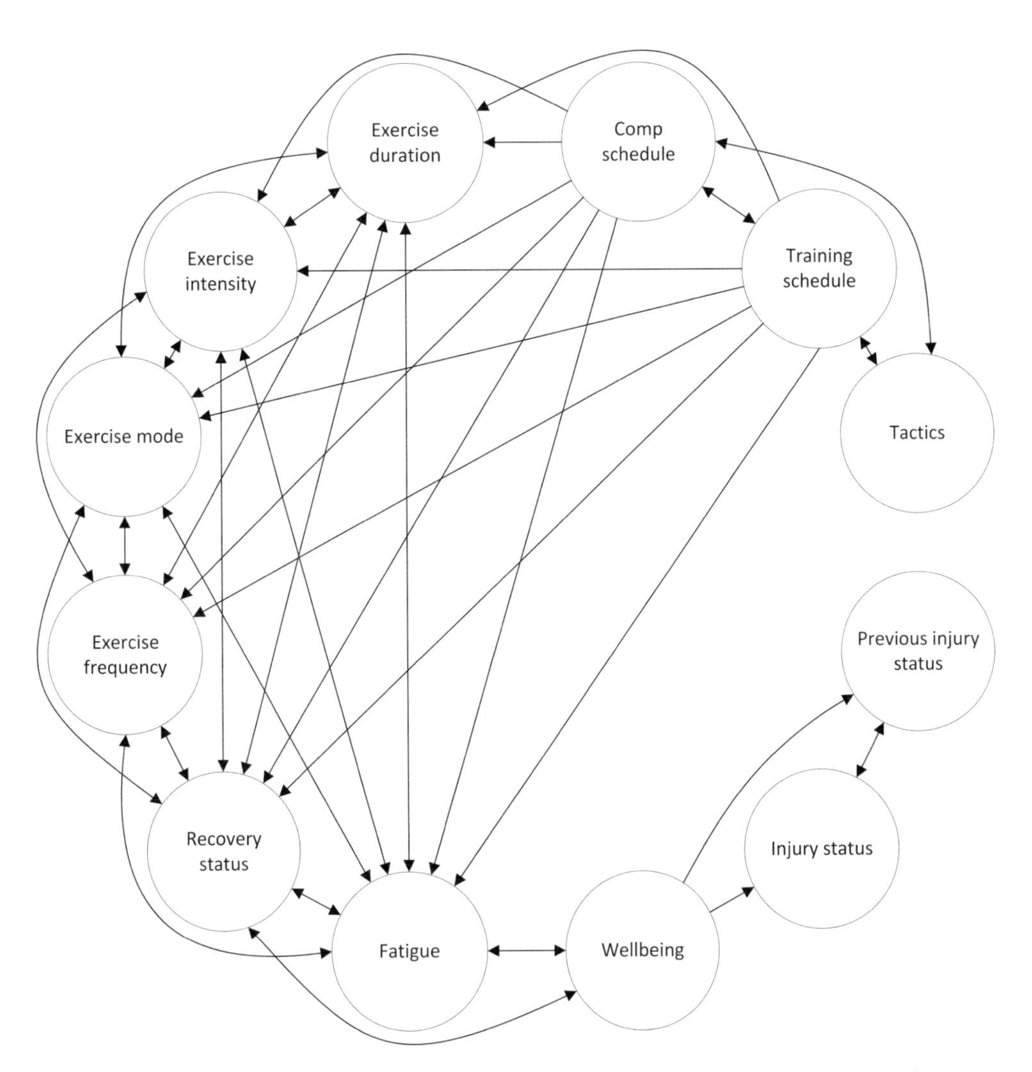

FIGURE 5.9 The network of information used by the actors in the social network that is required for player readiness Density (.42).

TABLE 5.12 Extract of task network (task to task) EAST-BL analysis

From (Task)	*To (Task)*	*Information not transferred*	*Outcome*	*Remedial measure*
Load management	S & C Training	Exercise intensity	Information about the intensity of exercise is not transferred, resulting in inappropriate strength and conditioning and injury and/or poor performance.	Employ wearable technology to capture real-time data on athletes' training intensity.
Health & wellbeing management	Recovery management	Recovery	Information on recovery is not transferred, resulting in poor recovery management and subsequent decline in health and wellbeing.	Hold regular meetings specifically focused on recovery status and strategies.
Technical/ tactical training	Education	Tactics	Information on match tactics is not transferred, resulting in players not being education on the way the coach wants to play, resulting in poor performance and lack of tactical progression.	Use apps or digital platforms where tactical plans can be shared and reviewed by players at their convenience.
Player monitoring	Manage player off-field	Wellbeing	Information on the players wellbeing is not transferred, resulting in inappropriate off-field management and potential occurrence of adverse incidents.	Develop individualised wellbeing plans for each player, taking into account their specific needs and circumstances.

TABLE 5.13 Extract of social network (agent to agent) EASTBL analysis

From (Agent)	*To (Agent)*	*Information not transferred*	*Outcome*	*Remedial measure*
Sports scientist	Head coach	Recovery	Coaches will not know how much recovery players have had, resulting in inappropriate training design with subsequent fatigued players and potential injury and/or poor performance.	Establish structured and consistent communication protocols for players to report their subjective feelings of recovery and fatigue.
Devices/ sensors (e.g., GPS)	Sports scientists	Exercise intensity	Sports scientists will not know exercise intensity, resulting in potential overload and injury, or under load and fitness losses and poor performance.	Ensure that data from devices and sensors is easily accessible, accurate, reliable, and regularly reviewed by coaching and medical staff.
Doctors	Coaches	Injury status	Coaches will not know how long a player is out injured for, resulting in inappropriate decision-making on player availability and/or transfers.	Enable coaches to access injury status information readily, so they are constantly informed about the player's rehabilitation status.
Coaches	Players	Tactics	Players will not know the match tactics, resulting in poor performances, and failure to improve.	Facilitate players to participate in the tactical planning process, allowing them to give input and ask questions to increase their understanding and buy-in.

Discussion

The EAST networks demonstrate the connectivity of the networks, and identify key tasks, actors, and information, both of which are critical to understanding DSA around the assessment of player readiness in a football club performance department.

Task networks

The task network shows a relatively high connectivity, via network density, indicating a tightly coupled and interdependent task network. This tight coupling has both positive and negative implications for player readiness. For example, a tightly coupled task network means that information, decisions, and actions related to readiness, e.g., S & C training, player monitoring, load management, and wellbeing assessments can be disseminated quickly throughout the network. This can be advantageous for comprehensive player management, ensuring that all aspects of a player's readiness are continuously monitored, and training and recovery is dynamically adjusted based on interconnected feedback. However, in a tightly coupled task network, if one task is not performed, performed inadequately, or disseminates incorrect information, it can rapidly and negatively impact the entire task network, and hinder the overall preparedness and performance of players. The network metrics revealed that 'education', and 'injury prevention' are the most prominent tasks in terms of their outgoing connections to other tasks, and their closeness to other tasks. This suggests that these tasks are key for influencing other player readiness tasks. 'Player monitoring' and 'health and wellbeing' were the most prominent tasks for in-degree centrality, indicating that many other tasks in the task network influence them. This shows the important role that all tasks play in maintaining player health and wellbeing.

Social network

The social network indicates a moderate level of connectedness between football performance department actors, suggesting that that while there is communication and collaboration among staff, there may be opportunities for further integration (McLean et al., 2021). For example, there might be opportunities to strengthen weak ties or to create new connections that can enhance collaboration, problem-solving, and decision-making processes. In the current social network, doctors, psychologists, and external trainers are not directly connected to the head coach, with these connections made though intermediatory sources such as the coaches, and sport scientists. As such, there may be opportunities for greater collaboration between these actors that could improve player readiness. While the results presented here are to demonstrate the EAST framework, there will be differences for individual clubs, and dependent on club resources. The key actors, across all network metrics were the players and the sports scientists. The players being a key actor in the social network with regard to player readiness is somewhat obvious, yet the central role of the sports scientist may be due to the breadth of roles that this position covers. For example, the sport scientist communicates with most of the other actors within the network, highlighting the importance of the role as both a gatekeeper and sign poster of critical information.

Information network

Based on density, the information network was not as tightly coupled as the task and social networks. One explanation for the relatively low density is that the same key pieces of information were connected frequently. For example, 'exercise duration', 'exercise intensity', 'exercise mode', 'exercise frequency', 'recovery', and 'fatigue' were identified as information that were highly connected within the information network. It is also common within EAST analyses that the information network is less connected than task and social networks, given that information can be specific to certain tasks and actors (McLean et al., 2023). According to the network metrics, 'training schedule', and 'match schedule' were key nodes within the information network. This is unsurprising given the importance on match preparation for matches and training. 'Recovery' and 'fatigue' were the pieces of information that contained the highest number of inbound connections, indicating that many other pieces of information are directly related to them. This is also unsurprising given that these represent multidimensional, latent constructs (Naughton et al., 2023) As mentioned, 'exercise duration', 'exercise intensity', 'exercise mode', and 'exercise frequency' all are highly connected to 'recovery' and 'fatigue'. This reflects the fact that these sources have the potential to influence recovery and fatigue in a dose-response relationship (Akubat, 2012).

EAST-BL

The EAST-BL analysis (Stanton & Harvey, 2017) demonstrated some of the potential risks when key pieces of information associated with player readiness are not transferred between tasks or between actors. While the current analysis demonstrated only a select few broken links and their outcomes, sports organisations could perform a full EAST-BL analysis to develop a comprehensive risk register that could be used to optimise player readiness. Indeed, using a method like EAST-BL to understand and identify potential risks (e.g., 'red teaming') is a valuable process that could lead to more effective teamwork, better decision-making within football departments, and improved outcomes for player readiness.

Recommended Reading

Hulme, A., McLean, S., Dallat, C., Walker, G. H., Waterson, P., Stanton, N. A., & Salmon, P. M. (2021). Systems thinking-based risk assessment methods applied to sports performance: A comparison of STPA, EAST-BL, and Net-HARMS in the context of elite women's road cycling. *Applied Ergonomics*, 91, 103297.

Salmon, P. M., Clacy, A., & Dallat, C. (2017). It's not all about the bike: Distributed situation awareness and teamwork in elite women's cycling teams. *Contemporary Ergonomics*, 240–248.

Neville, T., Salmon, P. M. (2016). Never blame the umpire – a review of Situation Awareness models and methods for examining the performance of officials in sport. *Ergonomics*, 59(7), 962–975.

References

Akubat, I. (2012). Training load monitoring in soccer: The dose-response relationships with fitness, recovery and fatigue (Doctoral dissertation, University of Hull).

Baber, C., McMaster, R.,... & Green, D. (2006). Distributed situation awareness in dynamic systems: theoretical development and application of an ergonomics methodology. *Ergonomics*, 49(12–13), 1288–1311.

Banks, V. A., & Stanton, N. A. (2019). Analysis of driver roles: Modelling the changing role of the driver in automated driving systems using EAST. *Theoretical Issues in Ergonomics Science*, 20(3), 284–300.

Guthrie, B., Jagim, A. R., & Jones, M. T. (2023). Ready or not, here I come: A scoping review of methods used to assess player readiness via indicators of neuromuscular function in football code athletes. *Strength and Conditioning Journal*, 45(1), 93–110.

Hulme, A., McLean, S., Dallat, C., Walker, G. H., Waterson, P., Stanton, N. A., & Salmon, P. M. (2021). Systems thinking-based risk assessment methods applied to sports performance: A comparison of STPA, EAST-BL, and Net-HARMS in the context of elite women's road cycling. *Applied Ergonomics*, 91, 103297.

Kirwan, B., & Ainsworth, L. K. (Eds.). (1992). *A guide to task analysis: The task analysis working group*. CRC Press. Boca Raton, FL.

Macquet, A. C., & Stanton, N. A. (2014). Do the coach and athlete have the same «picture» of the situation? Distributed Situation Awareness in an elite sport context. *Applied Ergonomics*, 45(3), 724–733.

McLean, S., King, B. J., Thompson, J., Carden, T., Stanton, N. A., Baber, C., Read, G. J. M., & Salmon, P. M. (2023). Forecasting Emergent Risks in Advanced AI systems: An Analysis of a Future Road Transport Management System., Ergonomics, doi:10.1080/00140139.2023.2286907

McLean, S., Rath, D., Lethlean, S., Hornsby, M., Gallagher, J., Anderson, D., & Salmon, P. M. (2021). With crisis comes opportunity: Redesigning performance departments of elite sports clubs for life after a global pandemic. *Frontiers in Psychology*, 11, 3728.

Mclean, S., Salmon, P.M., Gorman, A.D., Dodd, K. and Solomon, C. (2018). Integrating communication and passing networks in football using social network analysis. *Science and Medicine in Football*, 1–7.

Naughton, M., Scott, T., Weaving, D., Solomon, C., & McLean, S. (2023). Defining and quantifying fatigue in the rugby codes. *PLoS One*, 18(3), e0282390.

Neville, T. (2020). 10 Distributed Situation Awareness in Australian Rules Football Offciating. Human Factors and Ergonomics in Sport: Applications and Future Directions, 155.

Neville, T., Salmon, P. M. (2016). Never blame the umpire–a review of Situation Awareness models and methods for examining the performance of officials in sport. *Ergonomics*, 59(7), 962–975.

Osre, D., Wiggins, M., Williams, A., & Wong, W. (2000). Cognitive task analysis for decision centred design and training. *Task Analysis*, 170–190.

Roberts, A. P. J., Stanton, N. A., & Fay, D. T. (2017). Land Ahoy! – Understanding submarine command and control during the completion of inshore operations. *Human Factors*, 59(8), 1263–1288.

Salmon, P. M., Clacy, A., & Dallat, C. (2017). It's not all about the bike: Distributed situation awareness and teamwork in elite women's cycling teams. *Contemporary Ergonomics*, 240–248.

Salmon, P. M., Lenne, M. G., Walker, G. H., Stanton, N. A., & Filtness, A. (2014). Using the event analysis of systemic teamwork (EAST) to explore conflicts between different road user groups when making right hand turns at urban intersections. *Ergonomics*, 57(11), 1628–1642.

Salmon, P. M., & Plant, K. L. (2022). Distributed situation awareness: From awareness in individuals and teams to the awareness of technologies, sociotechnical systems, and societies. *Applied Ergonomics*, 98, 103599.

Salmon, P. M., Stanton, N. A., & Jenkins, D. P. (2017). Distributed situation awareness: Theory, measurement and application to teamwork. CRC Press. Boca Raton, FL.

Salmon, P. M., Stanton, N. A., Walker, G. H., Hulme, A., Goode, N., Thompson, J., & Read, G. J. (2022). *Handbook of systems thinking methods*. CRC Press. Boca Raton, FL.

Salmon, P. M., Walker, G. H., & Stanton, N. A. (2016). Pilot error versus sociotechnical systems failure? A distributed situation awareness analysis of Air France 447. *Theoretical Issues in Ergonomics Science*, 17(1), 64–79.

Stanton, N. A. (2014). Representing distributed cognition in complex systems: How a submarine returns to periscope depth. *Ergonomics*, 57(3), 403–418.

Stanton, N. A., & Harvey, C. (2017). Beyond human error taxonomies in assessment of risk in sociotechnical systems: A new paradigm with the EAST 'broken-links' approach. *Ergonomics*, 60(2), 221–233.

Stanton, N. A., Plant, K. L., Revell, K. M. A., Griffin, T. G. C., Moffat, S., & Stanton, M. J. (2019). Distributed cognition in aviation operations: A gate-to-gate study with implications for distributed crewing. *Ergonomics*, 62(2), 138–155.

Stanton, N. A., Salmon, P. M., Rafferty, L., Walker, G. H., Jenkins, D. P., & Baber, C. (2013). *Human factors methods: A practical guide for engineering and design*. 2nd Edition. Aldershot: Ashgate.

Stanton, N. A., Salmon, N. A., & Walker, G. H. (2018). *Systems thinking in practice: The event analysis of systemic teamwork*. Boca Raton, FL: CRC Press.

Stanton, N. A., & Young, M. S. (1999). What price ergonomics? *Nature*, 399(6733), 197–198.

Walker, G. H., Stanton, N. A., Baber, C., Wells, L., Gibson, H., Salmon, P. M., & Jenkins, D. P. (2010). From ethnography to the EAST method: A tractable approach for representing distributed cognition in air traffic control. *Ergonomics*, 53(2), 184–197.

Wasserman, S., & Faust, K. (1994). *Social network analysis: Methods and applications*. Cambridge: Cambridge University Press.

6
NETWORK ANALYSIS

Background

Network analysis is a powerful framework that enables us to study and analyse the intricate patterns of relationships that exist within social systems, including those in sport. It provides a set of theoretical concepts, analytical tools, and computational techniques to explore the complex interdependencies that shape our social world (Wasserman & Faust, 1994). By examining the connections between individuals, teams, organisations, network analysis uncovers hidden structures, identifies key actors, and reveals emergent properties that might not be apparent from individual-level analysis alone (Lusher et al., 2010; Wäsche et al., 2017). Consider a professional football team. Each player on the team is representative of a node within that network, and the communications and passes between them during a match represent the connections. By applying network analysis, analysts can reveal the most influential players who are central to the flow of the game, or to identify the overall playing style a team.

Social network analysis can be traced back to diverse disciplines such as sociology, anthropology, psychology, and mathematics. Later, the development of graph theory and mathematical modelling techniques by scholars provided the tools needed to quantify and analyse social networks mathematically (Freeman, 2004). In recent decades, the advent of computer technology and the explosion of digital data have revolutionised the field of social network analysis. Online social platforms, such as Facebook, X (formally Twitter), and LinkedIn, generate massive amounts of data that capture the interactions, connections, and communication patterns among millions of users. Likewise, in sport, advanced player tracking systems can be used to quantify the interactions between players in team. With data such as this, the possibilities are endless.

Given its simplicity and explanatory power, network analysis has numerous applications across a diverse range of sporting contexts.

DOI: 10.4324/9781003259473-8

Applications in sport

Within the past two decades network analysis has received increasing attention in sports research and practice (Palmer et al., 2023; Lord et al., 2022; Low et al., 2019). This is based on the realisation that sports teams and organisations are complex social systems that require coordination to achieve optimal performance (McLean et al., 2018). As such, there are multiple applications of network analysis in a sporting context. For example, passing networks for entire matches and/or different match phases (Clemente, 2015), goals scoring passing networks (McLean et al., 2017), intra-team communication (Lausic et al., 2009; 2015; McLean et al., 2018), broken passing networks (McLean & Salmon, 2019), and football player loans (Bond et al., 2020). Network analysis is typically applied to the analysis of invasions sports, such as football codes, basketball, water polo, hockey; however, it has been used in other sports such as golf, tennis, karate, swimming, and formula one (Wäsche et al., 2017). At the time of writing, the most popular application area of network analysis is football (soccer) (Lord et al., 2020).

Procedure and advice

A flowchart depicting the network analysis procedure is presented in Figure 6.1. Step-by-step guidance is presented below.

Step 1: Define analysis aims

First, the aims of the analysis should be clearly defined. As discussed earlier, there are many potential uses of network analysis in sport, including analysing team communications, team passing, athlete social networks, organisational structure, or the relationships between sports stakeholders such as club management and sponsors. For example, the aims of the analysis may be to evaluate passing connectivity between players during different match contexts, e.g., overall match, final third possession, or to identify key players in terms of passes made and received during the entire game. In addition, you may want to evaluate the communication of players during different phases of play, e.g., in possession (attacking) or not in possession (defending). Beyond this, you may be interested in a professional athlete support network in the run-up to a major event, or the communications occurring within an elite sports club. The analysis aims should be clearly defined so that appropriate scenarios are used, and relevant data are collected. Further, the aims of the analysis will dictate which network analysis metrics are used to analyse the networks (see Step 6).

Step 2: Collect data

The next step involves collecting the data that are to be used to construct the networks. Network analysis in sport can be applied from data from multiple sources, including on- and off-field platforms, e.g., player tracking technology through to social media. Given that passing is the most used data for network analysis in sport, the following guidance will focus on passing. However, there are a wide range of data sources for other types of networks, including social media, procedures, SME interviews, and communications logs to name only a few. Passing data can be collected for network analysis using various methods and technologies. Below are some common approaches.

Manual Event Tracking: This method involves manually recording passing events during a match. Trained observers or analysts watch the game and note down each pass,

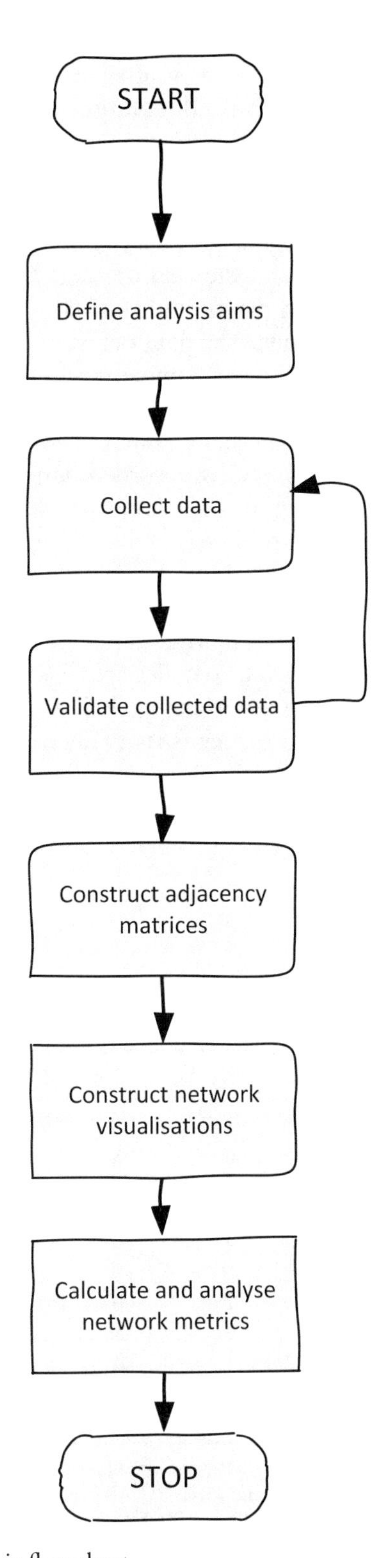

FIGURE 6.1 Network analysis flowchart.

including the passer, receiver, and any relevant attributes (e.g., pass direction, pass outcome). This approach is labour intensive and requires expertise in tracking game events accurately. It is often used in sports where automated tracking technologies are not available or affordable (sub-elite levels of sport).

Player Tracking Systems: The majority of professional sports now utilise player tracking systems that employ sensors, cameras, or GPS technology. These systems capture the positions and movements of players on the field or court throughout the game. Passing events can be inferred by analysing player movements, ball trajectories, or proximity between players during certain actions. The data collected from player tracking systems can provide detailed information about passing networks, including pass location, pass speed, pass accuracy, and player involvement.

Video Analysis: Video analysis is another common method for collecting passing data in sports. Sports analysts manually review recorded game footage and track passing events by visually identifying pass actions. This approach allows for a detailed examination of passing strategies, player positioning, and passing patterns. It can also involve annotating the passing events with additional attributes like pass type (e.g., short pass, long pass), pass quality, or pass outcome.

Wearable Technology: Some sports employ wearable technologies, such as smart jerseys or sensors embedded in equipment, to collect connectivity data. These devices can capture information related to player movements, ball possession, and passing interactions. The data collected can be used to reconstruct network dynamics.

It is now common for most sports technology companies to provide a network analysis module as part of their packages. However, if you are recording the interaction data (e.g., passing) yourself it is best practice to record passing interactions directly into a matrix (see Table 6.1).

TABLE 6.1 Example network matrix for passing connections in a football team

	GK	*RFB*	*RCD*	*LCD*	*LFB*	*RDMF*	*LDMF*	*RAMF*	*LAMF*	*RFWD*	*CFWD*
GK	0	3	3	5	3	1	0	0	0	5	1
RFB	3	0	4	1	0	6	6	2	0	11	0
RCD	3	5	0	9	2	4	11	1	0	1	0
LCD	5	1	8	0	5	1	16	2	0	0	0
LFB	2	0	0	4	0	4	14	2	4	1	1
RDMF	0	8	2	5	0	0	10	7	1	3	2
LDMF	0	6	5	6	17	7	0	17	11	7	3
RAMF	0	6	1	1	1	6	5	0	2	10	3
LFWD	0	1	1	1	3	1	3	0	0	1	1
RFWD	0	9	2	0	1	8	4	11	2	0	3
CFWD	0	1	1	0	0	0	2	2	0	1	0

Note: Values are the absolute number of passes between the intersecting positions during a match. For example, the RFB passed to the GK on three occasions, the RCD on four occasions, the LCD on one occasion, etc. Positions: GK– goalkeeper, RFB – right fullback, RCD – right central defender, LCD – left central defender, LFB – left fullback, RDMF – right defensive midfield, LDMF – left defensive midfield, AMF – attacking midfield, LFWD – left forward, RFWD – right forward, CFWD – central forward.

Step 3: Validate data collected

It is useful to revisit the data (if possible) to check for errors or missing data. This involves observing the task again (e.g., football match) and checking each connection was correctly recorded. It may also be pertinent for reliability purposes to get a second analyst to analyse a sub-set of the data in order to compute reliability statistics. If you are using data collected via a sports technology company, be sure to check for reliability and validity evidence. However, technology is now at a point where match-passing data is valid and reliable for evaluating match actions (Lui et al., 2013).

Step 4: Construct adjacency matrices

Once the data are checked and validated, the data analysis phase can begin. The first step involves the construction of an actor Adjacency matrix. This involves constructing a simple matrix and entering the frequency of connections (e.g., passes, communications) between each of the actors involved. For example, a simple adjacency matrix, showing the passing associations of a football team is presented in Table 6.1.

Step 5: Determine network characteristics

Understanding the distinctions between weighted, directed, and undirected networks prior to performing network analysis is essential, particularly in the context of sports. This enables the selection of the appropriate framework of metrics for analysis, ensuring that the interactions within a team or an athlete's performance network are represented accurately. For instance, weighted networks allow for a nuanced understanding of the strength or intensity of interactions, while directed networks are crucial for grasping the directional flow of these interactions, vital for tactical analyses in team sports (Lord et al., 2022; Ramos et al., 2018). Conversely, undirected networks offer a simplified view where the direction or weight of interactions is less relevant, such as in analysing team cohesion (Laporta et al., 2018). This understanding not only prevents misinterpretation of data but also guides strategic decision-making, allowing for the optimisation of both individual and team performance.

Weighted Network: In a weighted network, each edge (connection) is assigned a numerical weight or value that represents the strength, intensity, or some other quantitative attribute of the relationship between the connected nodes (Newman, 2004). The weights can indicate various aspects, depending on the context of the network. For example, in a social network, the weights could represent the frequency of communication or the level of influence between individuals (McLean et al., 2018). The inclusion of weights in the network allows for a more nuanced analysis, as it enables researchers to consider the varying degrees of interaction or importance between nodes.

Directed Network: In a directed network (also known as a 'digraph'), the edges have a specific direction, indicating the flow or asymmetry of the relationship between nodes (Ramos et al., 2018; Wäsche et al., 2017). The direction of an edge indicates that the relationship or interaction is unidirectional, from a source node to a target node (Clemente et al., 2016). This means that there can be different strengths or characteristics associated with each direction of the edge. Directed networks are useful in scenarios where the relationships are not symmetrical or where the flow of information, influence, or resources is important to understand.

Undirected Network: In an undirected network, the edges have no inherent direction (Wäsche et al., 2017). They represent symmetrical relationships between nodes, where the interaction or connection is bidirectional. In this type of network, the relationships between nodes are considered to be reciprocal or symmetric, without any specified source or target node (Wäsche et al., 2017). Undirected networks are commonly used when the relationships between nodes are based on mutual connections, such as friendship networks, collaboration networks, or co-occurrence networks.

Step 6: Construct social network visualisations

Next, network diagrams should be constructed (see below for network analysis programs). The social network diagram depicts each actor in the network and the connections that occurred between them during the scenario under analysis. Within the social network diagram, associations between actors are represented by directional arrows linking the actors involved. It is common to present the frequency of connections via the thickness of the arrows, e.g., a thicker arrow represents more connections than thinner arrows. Also, node size is commonly used for nodal prominence calculated via network metrics, e.g., a larger node represents prominent nodes for the specific network metrics being evaluated. An example passing network diagram for a football team is presented in Figure 6.2. Within Figure 6.2. the nodes (players) are positioned based on average field position derived from player tracking data.

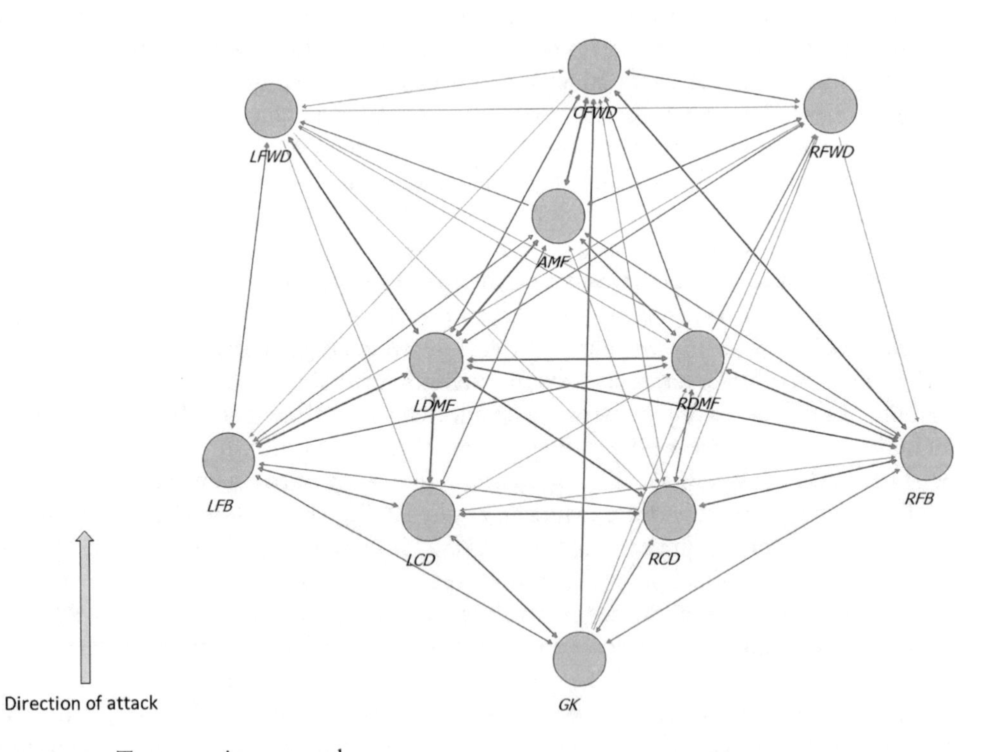

FIGURE 6.2 Team passing network.

Playing positions are based on average playing positions during the match. GK – goalkeeper, RFB – right fullback, RCD – right central defender, LCD – left central defender, LFB – left fullback, RDMF – right defensive midfield, LDMF – left defensive midfield, AMF – attacking midfield, RFWD – right forward, CFWD – central forward, LFWD – left forward.

Step 7: Calculate network metrics

Finally, the network should be analysed using appropriate network analysis metrics. Various metrics exist, and the metrics used are dependent on the analysis aims. For example, nodal metrics can be used to look at individual behaviour and status within the network, whereas overall network metrics can be used to look at the structure of the entire network. Table 6.2 presents a summary of common network metrics, their definitions, and their uses in sport. Table 6.2 is by no means an exhaustive list of network analysis metrics, but a brief guide to a selection of commonly applied metrics in sport.

TABLE 6.2 Network analysis metrics examples

Network metric	*Definition*	*Application in sport*
Network density	Network density calculates the proportion of actual connections (edges) in a network compared to the total possible connections	Network density can provide an overall measure of the connectedness within a team or sport. It helps assess the extent to which players or teams are connected through passing interactions
Out-degree centrality	Out-degree centrality is the count of outgoing connections that a node initiates towards other nodes in a network	Out-degree centrality can be used to identify players who frequently initiate passes or interactions with other players. It helps measure the proactivity, influence, or passing tendencies of a specific node within the network
In-degree centrality	In-degree centrality is the count of incoming connections that a node receives from other nodes in a network	In-degree centrality can be used to identify players who are frequently targeted or receive passes from other players. It helps measure the popularity, influence, or dependency on a specific node within the network
Betweenness centrality	Betweenness centrality quantifies the extent to which a node lies on the shortest paths between other nodes	Betweenness centrality can identify players that act as bridges, connecting different parts of the network. It helps identify key players who play a crucial role in passing the ball between different groups or areas on the field

(*Continued*)

TABLE 6.2 (Continued)

Network metric	*Definition*	*Application in sport*
Closeness centrality	Closeness centrality measures how quickly a node can access other nodes in the network	Closeness centrality can identify players that are in close proximity to other players in terms of passing. It helps identify players who have better access to the ball or have a higher likelihood of receiving passes quickly

Advantages

- Network analysis can be applied within sport to analyse a range of networks including team focused analyses focused on the communications and passing of players, identification of prominent players, interactions between match officials, player and official interactions with new technology and broader social and organisational networks such as athlete support networks, inter and intra-organisational communications, and the interactions between sport system stakeholders (e.g., clubs, fans, sponsors, and media).
- Network analysis is now commonly applied in sport to analyse network structures (Lord et al., 2022; Low et al., 2019; Palmer et al., 2023).
- Network diagrams provide a powerful and easily interpretable means of representing the connections between players.
- Highly suited to performance evaluation in a sporting context.
- Network analysis has been used in a range of domains for a number of different purposes.
- Network analysis is simple to learn and apply, and there are numerous online resources with guidance.
- Various free software programs are available for creating and analysing social networks.
- Network analysis is generic and can be applied in any domain in which associations between actors exist.

Disadvantages

- The data collection procedure for network analysis can be time-consuming, especially for large teams, if it is not automated.
- For complex collaborative tasks in which a large number of connections occur, network analysis outputs can become complex and time-consuming.
- Some knowledge of network statistics is required to understand the analysis outputs.
- Without the provision of software support, analysing the networks mathematically is a difficult and laborious procedure.

Approximate training and application times

The network analysis method requires a moderate level of training, with some knowledge of network statistics and metrics required to interpret the results. The application time is dependent upon the task and network involved. For short tasks involving small networks with only minimal associations between actors, the application time is low, particularly if a network analysis program is used. For tasks of a long duration involving large, complex networks, the application time is likely to be high, due to lengthy data collection and analysis processes. The application time is reduced dramatically via the use of software support that automates data collection, network diagram construction, and data analysis processes.

Tools or software needed

At a simplistic level, network analysis can be conducted using pen and paper only. However, it is recommended that computational programs be used to conduct network analysis using free online tools such as Gephi, R and igraph, SocNetV, and NodeXL, among many others.

Case study example: passing and communication networks of a professional soccer team

Introduction

Within the past decade, the acknowledgement of football (soccer) as a dynamic and complex system has brought about a rapid progression of new methods aimed at improving match performance analysis (Lord et al., 2022; Low et al., 2019; McLean et al., 2019a; Sarmento et al., 2017). As such, there has been a shift away from analysis of isolated components of football performance to a focus on group behaviours that explores the collective movements and cooperative actions of team members and opposition players (Low et al., 2019; Sarmento et al., 2017). Network analysis is now providing a detailed understanding of team functioning, and individual performance within the context of the team in soccer (Low et al., 2019; Sarmento et al., 2017; Wäsche et al., 2017). The use of network analysis in football match analysis has demonstrated how players interact via their passing connections, which reveals patterns of passing behaviour and identifies prominent players within the passing networks (Wäsche et al., 2017). When using network analysis to analyse passing connectivity in football, the players are seen as a network of nodes that are connected by passes (Grund, 2012).

Connectivity between team members in football does not, however, only relate to passing, players are also connected via verbal and non-verbal communication (Mclean et al., 2018). Team members use intra-team communication for information exchange during matches in order to optimise performance (Sullivan & Feltz, 2003). Furthermore, effective intra-team communication influences various team processes including motivation, concentration, strategy, skill acquisition, attitudes, and behaviour (Yukelson, 1993), and better performing teams communicate more, and use more task-related messages (Lausic et al., 2009).

The aim of this case study was to determine the playing positions of a professional football team that contribute the most to intra-team communication (ITC), and to analyse the passing contributions of the individual playing positions in competitive matches.

Design and procedure

The case study was based on data from a professional football team (n = 25 players) across a competitive 22 match season in the Australian A league. For the ITC data, each player completed a subjective rating scale following each match (see McLean et al., 2018). The rating scale required each player to rate the perceived amount of ITC received, and the perceived benefit of the received ITC to their match performances, from each of the other playing positions. The ITC rating scale ranged from 0 to 4, with 0 representing no benefit to performance, and 4 representing a very high benefit to performance. For the passing analysis, the match passing data were provided by Optasports, which is a reliable system for the analysis of match actions (Liu et al., 2013).

Network analysis

Network matrices were produced for ITC and passing data for each match (examples in Tables 6.3 and 6.4). The networks were directional (i.e., player A to player B) and indicated the strength of the connections between playing positions (i.e., the absolute values of the connections between players such as the total number of passes or level of beneficial ITC from player A and player B). For the current analysis, the determination of the inward and outward-bound connections was deemed important to understand the individual contributions of playing position to the beneficial ITC and passing. Therefore, the in-degree centrality (IDC) and out-degree centrality (ODC) were calculated for beneficial intra-team communication, and for passing using the network analysis program social network visualiser (SocNetV). The ODC represents the sum of connections from each player to all other players, and the IDC represents the sum of connections into each playing position from all other playing positions (Ribeiro et al., 2017). Only passes from members of the participating team were considered in the current analysis. For example, a regain of possession from a tackle or intercept was not considered as a contribution to IDC.

TABLE 6.3 Example social network matrix using the raw match passing data for one match

	GK	*RFB*	*RCD*	*LCD*	*LFB*	*RDMF*	*LDMF*	*RAMF*	*LAMF*	*RFWD*	*CFWD*
GK1	0	3	3	5	3	1	0	0	0	5	1
RFB2	3	0	4	1	0	6	6	2	0	11	0
RCD3	3	5	0	9	2	4	11	1	0	1	0
LCD4	5	1	8	0	5	1	16	2	0	0	0
LFB5	2	0	0	4	0	4	14	2	4	1	1
RMF6	0	8	2	5	0	0	10	7	1	3	2
CMF7	0	6	5	6	17	7	0	17	11	7	3
LMF8	0	6	1	1	1	6	5	0	2	10	3
LFWD9	0	1	1	1	3	1	3	0	0	1	1
CWD10	0	9	2	0	1	8	4	11	2	0	3
RFWD11	0	1	1	0	0	0	2	2	0	1	0

Values are the absolute number of passes between the intersecting positions during a match. For example, the GK passed the to the RFB on three occasions, the RCD on three occasions, the LCD on five occasions, etc.

TABLE 6.4 Example social network matrix using the intra-team communication data for one match

	GK	*RFB*	*RCD*	*LCD*	*LFB*	*RDMF*	*LDMF*	*RAMF*	*LAMF*	*RFWD*	*CFWD*
GK	0	4	4	4	4	2	3	4	4	0	0
RFB	2	0	3	3	2	2	2	2	3	3	3
RCD	3	3	0	4	3	2	3	2	3	3	3
LCD	1	1	3	0	3	2	1	3	0	0	0
LFB	3	0	0	4	0	4	1	2	0	1	2
RDMF	2	3	3	3	3	0	3	3	3	3	2
LDMF	3	3	3	4	3	3	0	3	3	3	3
RAMF	3	2	3	3	3	3	1	0	2	1	2
LAMF	1	4	2	1	0	4	3	2	0	1	1
RFWD	0	0	0	3	3	0	3	0	0	0	3
CFWD	1	2	2	3	2	2	3	3	2	2	0

Values are the perceived beneficial communication received during a match. For example, the RFB, RCD, LCD, and LFB perceived the communication from the GK was very highly beneficial to their performance in the match.

Results

The passing network is presented in Figure 6.3 and the network metrics in Table 6.5.

The communictaion network is presented in Figure 6.4 and the network metrics in Table 6.6.

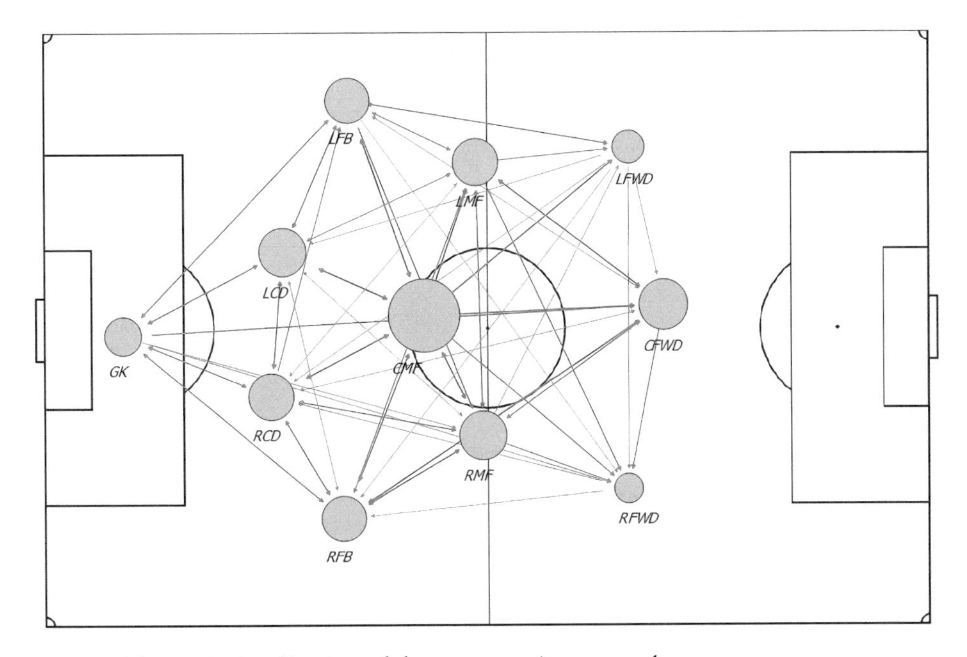

FIGURE 6.3 Network visualisation of the team passing network.

The node size is based on ODC, meaning that the node size is representative of the positions that provided the most passes to other positions. Passing network density was calculated as 0.76.

TABLE 6.5 Total passes and percentage ODC, and total passes received and percentage IDC metrics. The three highest values for each metric are shaded

Playing position	*Total passes*	*Percentage out-degree centrality*	*Total passes received*	*Percentage in-degree centrality*
GK	21	5.6	13	3.5
RFB	33	8.8	40	10.7
RCD	36	9.7	27	7.2
LCD	38	10.2	32	8.6
LFB	32	8.6	32	8.6
RMF	38	10.2	38	10.2
CMF	79	21.3	71	19.1
LMF	35	9.4	44	11.8
LFWD	12	3.2	20	5.3
CWD	40	10.7	40	10.7
RFWD	7	1.9	14	3.8

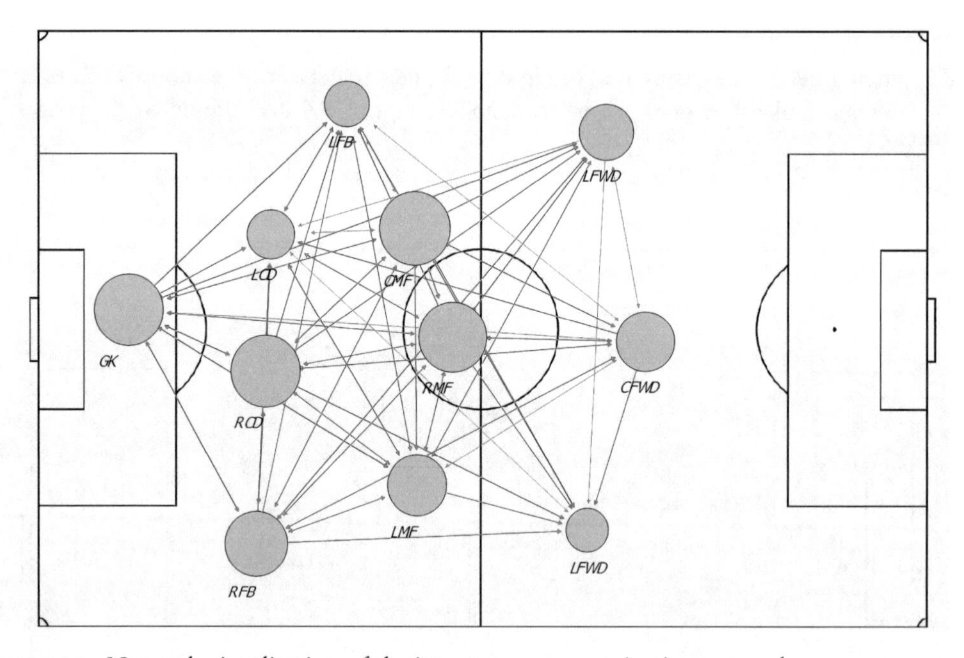

FIGURE 6.4 Network visualisation of the intra-team communication network.

The node size is based on ODC, meaning that the node size is representative of the positions that provided the most beneficial communication to other positions.

TABLE 6.6 Percentage ODC, and percentage IDC metrics for beneficial intra-team communication. The three highest values for each metric are shaded

Playing position	*Percentage out-degree centrality*	*Percentage in-degree centrality*
GK	11.8	7.7
RFB	10.2	8.9
RCD	11.8	9.3
LCD	6.1	13.1
LFB	5.3	10.6
RMF	11.4	8.1
CMF	12.6	9.3
LMF	9.3	9.7
LFWD	7.7	8.1
CWD	8.9	7.3
RFWD	4.9	7.7

Discussion

This case study investigated, using network analysis, the contributions of ITC and passing by individual playing positions in competitive professional football. Overall, the density of the passing network was .76, which indicates that multiple players are connected to one other via passes. The left central defender, central defensive midfielder, and central forward positions were the most prominent contributors to the passing network. The right fullback, central midfielder, left midfielder, and centre forward received the highest number of passes. For ITC, the goalkeeper, right central defender, and central midfield positions were perceived to provide the communications that most benefited performance, whereas the left central defender, left fullback, and left midfield positions perceived the communication they received be beneficial to their performance.

The passing analysis in this study supports previous analyses, whereby the central defenders and central defensive midfield positions contributed the highest values for passing compared to the other playing positions within the team (Clemente et al., 2014). However, the high IDC and ODC metrics for the centre forward are rare and offer potential insights into the team's playing style. These findings, together with the high density of the passing network indicate that the team under analysis, played a possession-based style with the central spine of the team contributing the majority of passing. For an opposition analyst with this information, it provides opportunities to devise tactics to potentially disrupt their preferred playing style. For example, it may be beneficial to overload the central areas and force them to play wider and cut off the passing options to the centre forward. These findings also provide the analysts and coaches from the team under analysis to assess if they are achieving their planned match tactics regarding passing.

For the ITC, the findings provide valuable information on who provides and receives the most beneficial communication. Similar to the passing metrics, it was the central spine of the team that contributed the most beneficial communication, indicating that these positions are key communication distributors among the team, which has implications for team composition. Understanding the communication processes of a football team has several other practical implications. For example, knowing who can effectively communicate game plans is crucial for coaches, teaming good communicators with youth players in training will likely assist in player development, and measuring ITC may assist in talent identification. This case study application of network analysis has provided a snapshot of team functioning and the connectivity among team members that can guide coaches' decision-making for training design, team tactics, and for team selections. While this case study analysis was of one match, more powerful analyses can be obtained through aggregation of matches for a season, a competition/tournament, for matches against top teams and lower teams, periods of fixture congestion, among many other contexts.

Recommended Reading

Wäsche, H., Dickson, G., Woll, A., & Brandes, U. (2017). Social network analysis in sport research: An emerging paradigm. *European Journal Sport Society*, 14(2), 138–165.

Lusher, D., Robins, G., & Kremer, P. (2010). The application of social network analysis to team sports. *Measurement in Physical Education and Exercise Science*, 14(4), 211–224.

Mclean, S., Salmon, P. M., Gorman, A. D., Dodd, K., & Solomon, C. (2018c). Integrating 1communication and passing networks in football using social network analysis. *Science and Medicine Football*, 3(1), 29–35.

Clemente, F. M., Martins, F. M. L., Kalamaras, D., Wong, D. P., & Mendes, R. S. (2015). General network analysis of national soccer teams in FIFA World Cup 2014. *International Journal of Performance Analysis in Sport*, 15(1), 80–96.

References

Bond, A. J., Widdop, P., & Parnell, D. (2020). Topological network properties of the European football loan system. *European Sport Management Quarterly*, 20(5), 655–678.

Clemente, F. M., Martins, F. M. L., Kalamaras, D., Wong, D. P., & Mendes, R. S. (2015). General network analysis of national soccer teams in FIFA World Cup 2014. *International Journal of Performance Analysis in Sport*, 15(1), 80–96.

Clemente, F. M., Martins, F. M. L., Wong, D. P., Kalamaras, D., & Mendes, R. S. (2015). Midfielder as the prominent participant in the building attack: A network analysis of national teams in FIFA World Cup 2014. *International Journal of Performance Analysis in Sport*, 15(2), 704–722.

Clemente, F. M., Silva, F., Martins, F. M. L., Kalamaras, D., & Mendes, R. S. (2016). Performance analysis tool for network analysis on team sports: A case study of FIFA Soccer World Cup 2014. *Proceedings of the Institution of Mechanical Engineers, Part P: Journal of Sports Engineering and Technology*, 230(3), 158–170.

Freeman, L. C. (2004). *The development of social network analysis: A study in the sociology of science*. Vancouver, BC: Empirical Press.

Grund, T. U. (2011). Network structure and team performance: The case of English Premier League soccer teams. *Social Network*, 34(4), 682–690.

Laporta, L., Afonso, J., & Mesquita, I. (2018). The need for weighting indirect connections between game variables: Social Network Analysis and eigenvector centrality applied to high-level men's volleyball. *International Journal of Performance Analysis in Sport*, 18(6), 1067–1077.

Lausic, D., Tennebaum, G., Eccles, D., Jeong, A., & Johnson, T. (2009). Intrateam communication and performance in doubles tennis. *Research Quarterly for Exercise and Sport*, 80(2), 281–290.

Lausic, D., Razon, S., & Tenenbaum, G. (2015). Nonverbal sensitivity, verbal communication, and team coordination in tennis doubles. *International Journal of Sport and Exercise Psychology*, 13(4), 398–414.

Lord, F., Pyne, D. B., Welvaert, M., & Mara, J. K. (2022). Field hockey from the performance analyst's perspective: A systematic review. *International Journal of Sports Science & Coaching*, 17(1), 220–232.

Low, B., Coutinho, D., Gonçalves, B., Rein, R., Memmert, D., & Sampaio, J. A. (2019). Systematic review of collective tactical behaviours in football using positional data. *Sports Medicine*, 50, 343–385.

Liu, H., Hopkins, W., & Gómez, A. M. (2013). Molinuevo. Inter-operator reliability of live football match statistics from OPTA Sportsdata. *International Journal of Performance Analysis in Sport*, 13(3), 803–821.

Lusher, D., Robins, G., & Kremer, P. (2010). The application of social network analysis to team sports. *Measurement in Physical Education and Exercise Science*, 14(4), 211–224.

McLean, S., Read, G. J. M., Hulme, A., Dodd, K., Gorman, A. D., Solomon, C., & Salmon, P. M. (2019a). Beyond the tip of the iceberg: Using systems archetypes to understand common and recurring issues in sports coaching. *Frontiers in Sports and Active Living*, 1, 49.

McLean, S., & Salmon, P. M. (2019b). The weakest link: A novel use of network analysis for the broken passing links in football. *Science Medicine in Football*, 3(3), 255-258..

McLean, S., Salmon, P. M., Gorman, A. D., Stevens, N. J., & Solomon, C. (2018a). A social network analysis of the goal scoring passing networks of the 2016 European Football Championships. *Human Movement Science*, 57, 400–4008.

McLean, S., Salmon, P. M., Gorman, A. D., Wickham, J., Berber, E., & Solomon, C. (2018b). The effect of playing formation on the passing network characteristics of a professional football team. *Human Movement*, 5, 14–22.

Mclean, S., Salmon, P. M., Gorman, A. D., Dodd, K., & Solomon, C. (2018c). Integrating 1communication and passing networks in football using social network analysis. *Science and Medicine Football*, 3(1), 29–35.

McLean, S., Salmon, P. M., Gorman, A. D., Naughton, M., & Solomon, C. (2017a). Do intercontinental playing styles exist? Using social network analysis to compare goals from the 2016 EURO and COPA football tournaments knock-out stages. *The Theoretical Issues in Ergonomics Science*, 18(4), 370–383. doi:10.1080/1463922X.2017.1290158

McLean, S., Salmon, P. M., Gorman, A. D., Read, G. J. M., & Solomon, C. (2017b). What's in a game? A systems approach to enhancing performance analysis in football. *PLoS One*, 12(2), e0172565.

Newman, M. E. (2004). Analysis of weighted networks. *Physical Review E*, 70(5), 056131.

Palmer, S., Novak, A. R., Tribolet, R., Watsford, M. L., & Fransen, J. (2023). Cooperative networks in team invasion games: A systematic mapping review. *International Journal of Sports Science & Coaching*, 18(6), 2347–2359.

Ramos, J., Lopes, R. J., & Araújo, D. (2018). What's next in complex networks? Capturing the concept of attacking play in invasive team sports. *Sports Medicine*, 48, 17–28 & Coaching, 17479541231177133.

Sarmento, H., Clemente, F. M., Araújo, D., Davids, K., McRobert, A., & Figueiredo, A. (2017). What performance analysts need to know about research trends in association football (2012–2016): A systematic review. *Sports Medicine*, 48(4), 799–836.

Sullivan, P., & Feltz, D. L. (2003). The preliminary development of the scale for effective communication in team sports (SECTS). *Journal of Applied Social Psychology*, 33(8), 1693–1715,

Wäsche, H., Dickson, G., Woll, A., & Brandes, U. (2017). Social network analysis in sport research: An emerging paradigm. *The European Journal for Sport and Society*, 14(2), 138–165.

Wasserman, S., & Faust, K. (1994). *Social network analysis: Methods and applications*. Cambridge: Cambridge University Press.

Yukelson D. (1993). Communicating effectively. *Applied Sport Psychology: Personal Growth to Peak Performance*, 122–136.

7

CAUSAL LOOP DIAGRAMS

Background

In a world increasingly obsessed with data and quantification, it is tempting to reduce complex systems to individual, isolated, and measurable variables. However, individual variables alone often fail to capture the intricate and dynamic web of activities that constitute any system – especially one as complex and multifaceted as sport. As athletes, coaches, and researchers alike seek a deeper understanding of performance dynamics, appropriate tools are required. One such tool is Causal Loop Diagrams (CLDs).

All systems comprise interacting networks of positive and negative feedback loops that influence behaviour (Sterman, 2000). CLDs offer a means of representing these feedback loops and interdependencies. Originating from the field of system dynamics, CLDs have been applied to various complex problems ranging from the transmission of COVID-19 (Kumar et al., 2021; Sahin et al., 2020), major accidents (e.g., Lin & Chien, 2019), and obesity (Allender et al., 2015) to issues impacting research quality and translation (McLean et al., 2021), racism (Burrell et al., 2021), transport safety (Shepherd, 2014) and climate change (Richards et al., 2021). CLDs help us grasp how variables not only interact, but also feed back into each other in a self-perpetuating cycle – either virtuous or vicious (Meadows, 2008).

CLDs comprise relevant variables connected by arrows which depict the causal influences between them (Sterman, 2000). Reinforcing loops (or positive feedback) are actions that show how change in one variable produces change in another variable in the same direction, e.g., births increase the population which further increases births (Figure 7.1). Balancing loops (or negative feedback), however, work to keep the system in a state of equilibrium (Kim, 1993; Senge, 1990). Balancing loops resist change in one direction by producing change in the opposite direction. For example, an increasing population will lead to an increased death rate which in turn acts as a balance on population growth (Figure 7.1).

DOI: 10.4324/9781003259473-9

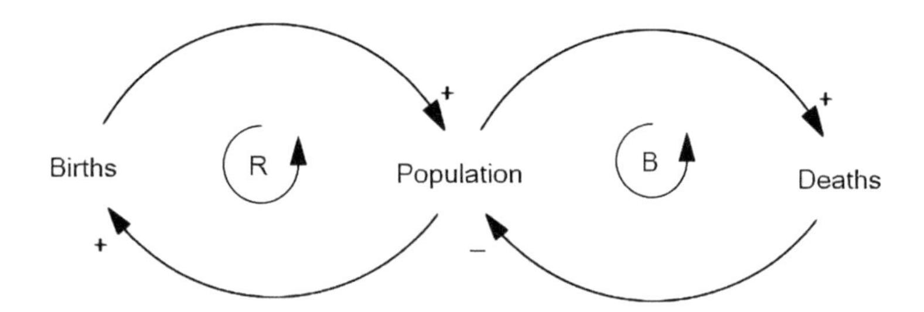

FIGURE 7.1 CLDs representing the reinforcing loop (R) of births on the population, and the balancing loop (B) of deaths on the population. + indicates an increasing (or positive) effect on the variable, and – indicates a balancing (or negative) influence on the variable.

Applications in sport

Causal loop diagrams have been applied in various contexts in sport. For example, Salmon et al, (2021) applied CLD to understand the key variables that interact to influence elite football club performance. CLDs have also been applied to understand the interaction of systemic factors that influence the research practice gap in sports science research (McLean et al, 2021). Naughton et al, (2024) investigated the positive and negative impacts of technology insertion in sport through CLDs. CLDs have been developed for recreational running injury and subsequently used to inform computational models of running injury though system dynamics (Hulme et al, 2018). The flexibility of CLDs allow them to be applied across a broad range of topics and issues in sport.

Procedure and advice

The CLD procedure is shown below in Figure 7.2, and the main structural elements of CLDs are presented in Figure 7.3. CLDs are typically developed using a group model-building process (Berard, 2010; Hovmand, 2014; Sterman, 2000). Berard (2010), and Salmon et al (2020) describe two forms of group model building. The first involves the modellers themselves constructing the CLD using data derived from sources such as documentation review, interviews, and research study findings. The second form involves engaging relevant SMEs to assist in building the models. It is often useful to combine these approaches, with initial development of the model based on data, followed by SMEs being involved in the review and refinement of the model (McLean et al., 2021).

Step 1: Define aims and focus of the analysis

The first step in applying CLDs involves clearly defining the aims and boundaries of the analysis. The CLD presented in the case study example in this chapter was developed to examine the potential impact of introducing artificial intelligence (AI) technologies in sport.

Step 2: Pre-group modelling data collection

Once the issue under analysis and aims of the analysis are clearly defined, specific data regarding the issue and system should be collected. The data collection activities

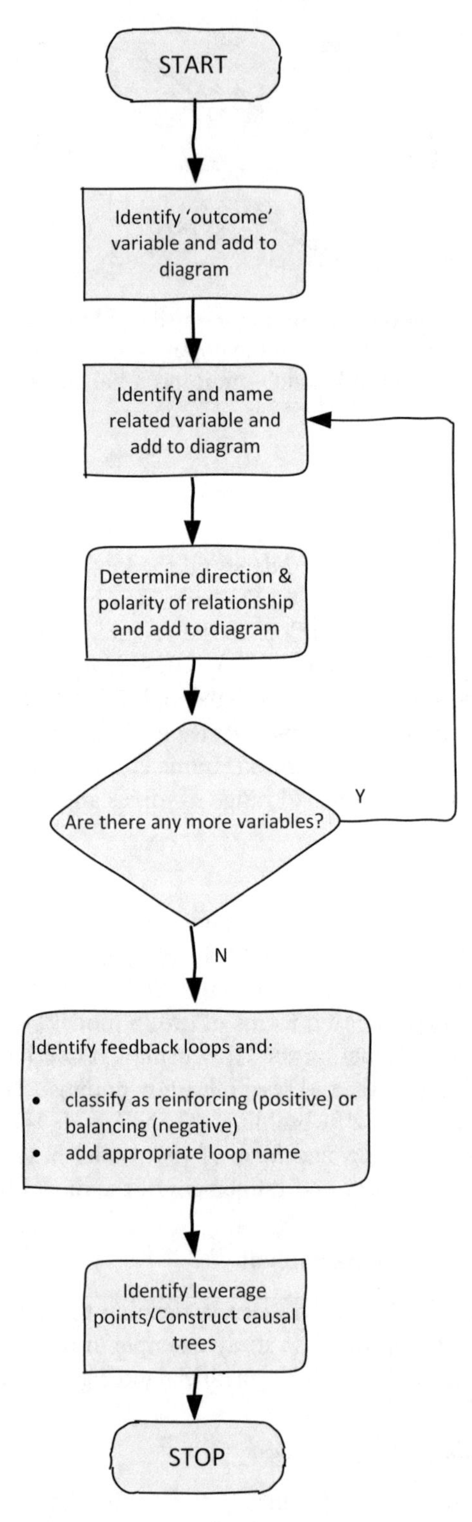

FIGURE 7.2 CLD flowchart.

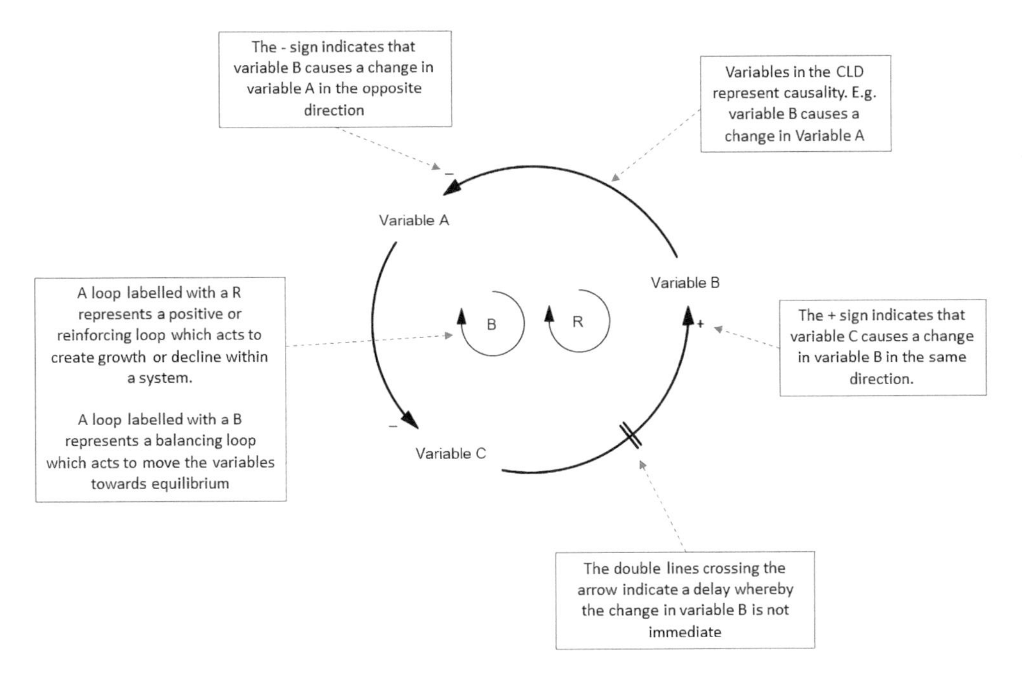

FIGURE 7.3 CLD elements and descriptions.

undertaken are dependent upon the aims and boundaries of the analysis as well as project constraints, such as time, number of analysts available, and access to information and personnel. Data to support CLD development can be collected from a variety of sources, with the focus on gathering information on the known variables that influence behaviour. Previous CLDs in sport have utilised peer-reviewed literature, and documentation review, and SME workshops and/or interviews (McLean et al., 2021).

Step 3: Identify and name variables

Steps 3–6 can either be undertaken by the analyst team or directly with SMEs as part of a workshop process. It is recommended, however, that Steps 3–6 be undertaken initially by the analyst team and then SMEs are used during Steps 7 and 8 to review and refine the CLD.

Step 3 involves identifying and naming the different variables that interact to influence the behaviour or problem of interest. It can be useful at this stage to begin with the outcome variables. For example, in the case study example, construction of the CLD began with identifying variables related to AI implementation in sport, such as 'injury/playing time lost' or 'adoption/dependence'. When identifying and naming variables de Pinho (2015) provides the following advice:

- Use nouns for variable names and try to avoid the use of verbs, e.g., 'efficiency' is more appropriate than 'improve productivity'.
- Variable names should be neutral in terms of direction, e.g., 'accuracy' should be used and not 'high accuracy' or 'low accuracy'.

- Variables should be measurable and observable and can either increase or decrease over time.
- Where possible the positive instantiation of the variable should be used rather than the negative, e.g., 'quality of performance' rather than 'poor performance'.

It is often useful when identifying variables to use scenarios, case studies, or stories to help the group think about influences on behaviour.

Step 4: Determine and record relationships between the variables

Step 4 can occur in conjunction with Step 3, or following Step 3. Once the key variables are identified, the next step involves identifying how the variables affect one another and marking this on the CLD. This involves the group discussing each variable and its relationships with other variables in the CLD. Where a relationship is specified, a directional link should be added to the CLD. For example, for the AI implementation in sport case study example later, a relationship was identified between 'quality of data collection and analysis' and 'efficiency', as it was determined that the quality of data collection and analysis would influence the efficiency of AI tools (see Figure 7.4). The polarity of the links should also be performed at this point. For example, a + indicates a change in the same direction, whereas, a – indicates a change in the opposite direction (Figure 7.3).

Delays should also be added to relationships where the effect of the cause is not immediate and takes time to occur. For the case study example, a delay was added to the relationship between 'dependence' and 'skill/creativity' as the increase in creativity and skill degradation is not immediate and can take some time to occur. This is shown in Figure 7.5.

New variables are also likely to be found during this step when the group identifies relationships with variables not already identified. In this case, the new variables and identified relationships should also be added to the CLD. In addition to identifying new variables, undertaking this step also prompts the group to refine and validate variable names or group variables, where possible, to create a more parsimonious and valid model.

Steps 3 and 4 should continue until there is group consensus that all relevant variables and relationships have been identified and added to the draft CLD.

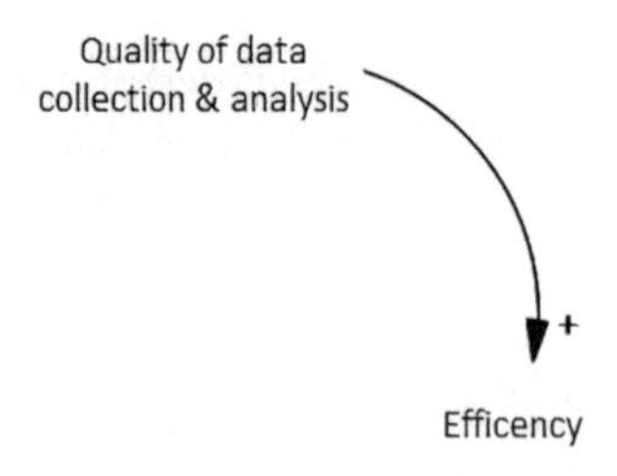

FIGURE 7.4 Adding relationships example. The + indicates that 'efficiency' increases with 'quality of data collection and analysis'. Alternatively, the + also indicates that a reduction in 'efficiency' will result from a reduction in 'quality of data collection and analysis' (i.e., the + indicates an effect in the same direction of the cause).

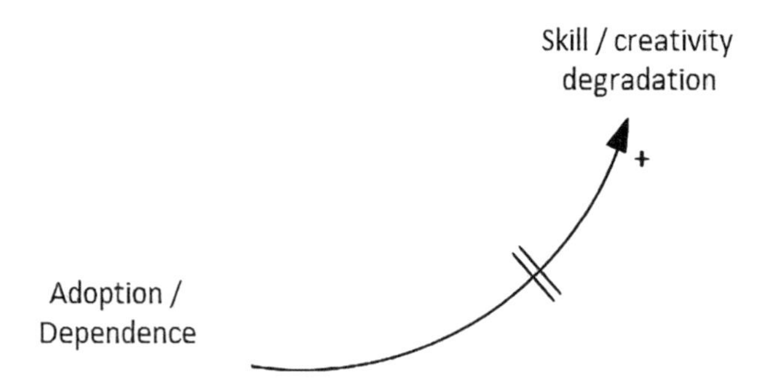

FIGURE 7.5 Delay demarcation. As the time period between the dependence on technology and skill/creativity degradation of humans is not immediate, the delay demarcation is used to indicate this.

Step 5: Classify and name loops

Once all relevant variables and relationships have been added to the CLD, the next step involves identifying key causal loops and classifying them as either reinforcing (positive) loops or balancing (negative) loops. Each identified loop should also be named at this point. For the case study example, the 'adoption/dependence of technology in sport' loop presented in Figure 7.6 is a balancing loop that shows how an increased adoption of, and dependence on, technology results in a degradation of skill/creativity of analysts and other performance staff, which in turn causes decreased satisfaction at work which acts to balance adoption of and dependence on technology.

The identification and naming of pertinent causal loops can also be supported by the creation of behaviour over time graphs or reference modes. These do not need to be data-driven at this stage and can be developed based on the group's perceptions regarding the behaviour of key variables over time. For the technology implementation in sport case study example, a behaviour over time graph would represent Panel C (see Figure 7.7) which indicates oscillations in the adoption and dependence of technology implementation in sport. Panels A and B (see Figure 7.7) both represent reinforcing loops that drive behaviour in a certain direction.

Step 6: Build CLD in software tool

Once the draft CLD is complete, the next step involves building the CLD within a suitable software package, such as Vensim or Stella Architect (assuming it was not developed initially within a software program). Further refinement can also occur during this process. It is recommended that the CLD diagram is formatted in a manner that makes it easy to interpret. For example, the analyst should attempt to avoid arrows crossing over one another, to provide sufficient space between variable names, and ensure all polarity markings are clearly presented. It should be noted that Steps 3–6 can occur without the use of specific software, such as using pen and paper, whiteboards, and post-it notes.

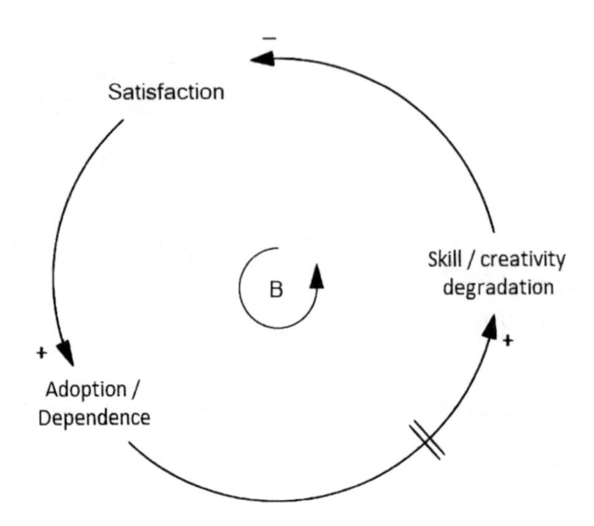

FIGURE 7.6 The 'adoption/dependence of technology in sport' loop.

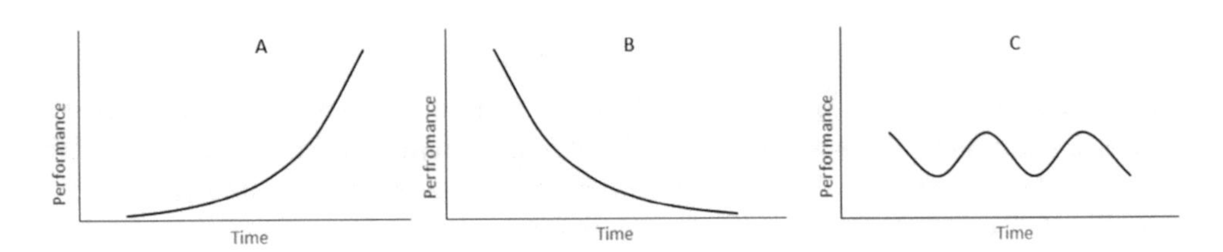

FIGURE 7.7 Conceptual behaviour over time graphs. Panels A and B represent reinforcing feedback loops, and Panel C represents a balancing feedback loop.

Step 7: Review and refine analysis

SME review is a critical component of CLD development. Once the initial draft CLD is complete it is useful to have various SMEs review it, including those involved in the development of CLD and others external to the process. SME review can be undertaken in various ways, including workshops, focus groups, Delphi study, or one-on-one sessions. It is recommended that the SMEs are first given a description of the CLD, including all variables, relationships, and positive and negative feedback loops. Next, the SMEs should be asked to review each variable and relationship and comment on variables names, relationships and their direction, feedback loop names, and also whether any variables or relationships are missing from the CLD. It is useful during this stage to have a data dictionary which provides a clear definition of each variable and relationship within the CLD. The CLD should then be refined based on the SME's feedback. This typically involves adding new variables and relationships or modifying variable and/or loop names.

Step 8: Identify leverage points

CLDs are useful in that they support the identification of leverage points (see Chapter 2) that reflect places in the system where small interventions could have significant impacts on system behaviour (Meadows, 2008). If required, the analyst team and SMEs should review the CLD and seek to identify leverage points. This can be supported via software tools' construction of causal trees which show the incoming and outgoing relationships from individual variables. Causal trees can also be a useful method for demonstrating the systemic influence of key variables within CLD. Causal trees demonstrate, through direct and indirect links, the systemic influence of variables across the system on individual variables through primary and secondary links (see Mclean et al, 2021).

Advantages

- The output provides an intuitive model of the causal structure underlying system behaviour.
- The output provides a comprehensive overview of the variables that interact to create the issue under analysis.
- Polarity and delay markings showing cause directionality and temporality add an additional layer to the analysis.
- Causal trees and network analysis metrics can help in the interpretation of the CLD.
- CLDs are generic and can be applied to any issue in any domain.
- CLDs are highly flexible and can be used with varying degrees of granularity and scope.
- Initial CLD development can be quick, noting that further work is required as part of a group modelling process.
- CLDs are straightforward to apply and require minimal training.
- Various software programs are available to support the development of CLDs, including Vensim and Stella Architect.
- CLDs can be used to inform the development of quantitative system dynamics models which enable simulation of system behaviour over time.

Disadvantages

- The quality of the CLD is highly dependent on the involvement of multiple SMEs during development.
- Initial data collection activities and group modelling processes can be time-consuming.
- CLDs can become overly complex, large, and difficult to interpret.
- It is possible to create erroneous or 'faulty' models with incorrect assumptions about causality and system dynamics (Mirchi et al., 2012).
- Unless converted into system dynamics models, CLDs are difficult to quantitatively test.
- There is no reliability and validity evidence associated with CLDs.

Related methods

CLDs are often used to inform the development of quantitative system dynamics models. Specifically, the CLD is used to identify the positive and negative feedback loops which define the functional architecture of the system (Sterman, 2000). The key loops within the CLD are then articulated as formal dynamic hypotheses, and the overall architecture of the system is captured in a quantitative system dynamics model. The system dynamics model comprises stocks, flows, and information (auxiliary) variables and can be used to simulate system behaviour over time. Important in this conversion, however, is an appreciation that units of measurement within CLDs are ill-defined. By contrast, units of measurement in system dynamics models must be explicitly defined and conserved throughout the operation of the model.

Approximate training and application times

CLDs are a relatively simple method that requires little training, and various courses are available online. Application time depends on the system or issue under analysis; however, CLD analyses typically require significant time as they require initial data collection activities and multiple model iterations. This may require multiple group modelling sessions with SMEs to refine the CLD. They can, however, also be entirely conceptual, requiring little to no 'real-world' data to either create or qualitatively validate among model builders and stakeholders.

Tools needed

CLDs can be created using pen and paper or a whiteboard and markers; however, once a draft model is developed it is useful to use a software program such as VenSim or Stella Architect to create the final CLD. Further, CLD tools have additional features such as causal trees that can be used to interpret the model.

Case study example: advanced technology implementation in sport

Background

Sport systems are becoming increasingly reliant on advanced technologies (Chmait & Westerbeek, 2021). Given the substantial financial rewards associated with success in sport, athletes, clubs, leagues, numerous sporting organisations are adopting AI in pursuit of obtaining a competitive advantage. A key benefit of AI is their capacity to assimilate and interpret large volumes of data to provide insight into sports performance, which was previously the role of humans (Araújo et al., 2021). Research using AI in sport has included the development and evaluation of athlete performance measures, the prediction of match outcome and injury, match analysis, and the development of training programmes for athletes, to name a few (Bunker & Susnjak, 2022; Cust et al., 2019; Herold et al., 2019; Van Eetvelde et al., 2021). However, the need to jointly optimise the human and technological elements of sport systems is perhaps more pressing than it has ever been, and this will only increase with further technological innovations (Endsley, 2023). Though there are potentially widespread benefits, the risks associated with poorly designed AI have been discussed in many areas (Hancock, 2017, 2022; McLean et al., 2023; Müller, 2016; Omohundro, 2013; Salmon et al., 2023). These risks are varied and range from safety and privacy risks to ethical and even existential risks (McLean et al., 2021; Müller, 2016). It is therefore critical to examine the potential impacts of introducing AI in sport. The aim of this case study was to use CLDs to model the feedback loops that could create both positive and negative impacts when implementing AI in sport.

Method

In the present case study, a CLD-based approach was conducted to explore and model the dynamics of the implementation of AI in sport. The CLD was developed through a collaborative group model-building process (Sterman, 2014). The authors utilised relevant peer-reviewed literature and their own subject matter expertise in sport science, systems modelling in sport, risks associated with advanced technologies, and systems thinking (e.g., McLean et al., 2021; Salmon et al., 2022). The models were produced using Vensim software (Ventana Systems Inc, Harvard, MA).

Results

The developed CLD for AI technology implementation in sport is presented in Figure 7.8.

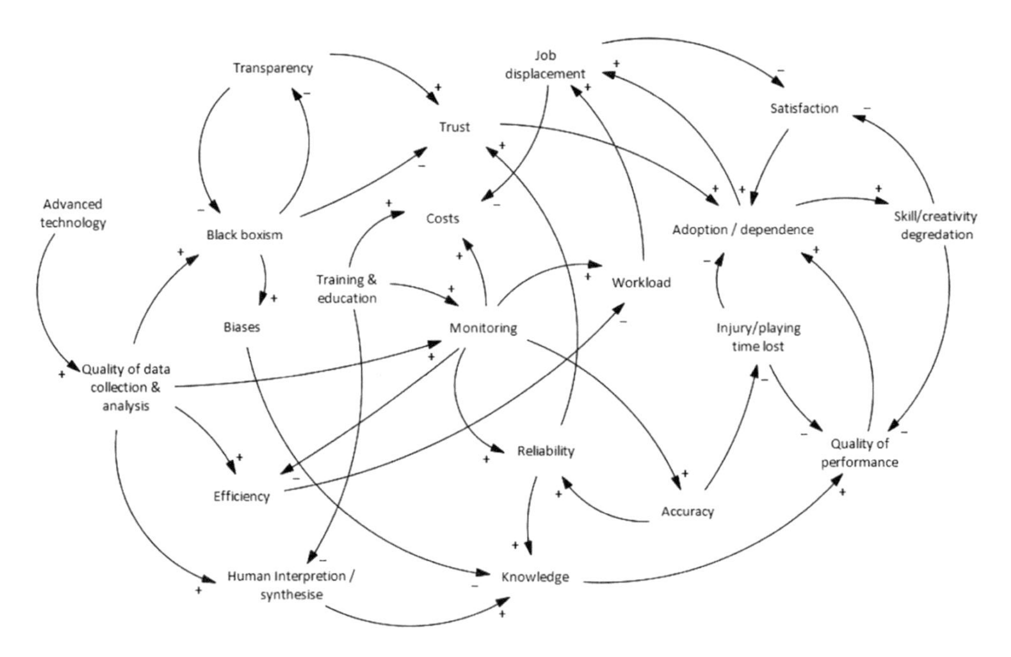

FIGURE 7.8 CLD of advanced technology implementation in sport. Adapted from Naughton et al, 2024.

Discussion

The CLD demonstrates the inherent complexity of advanced technology implementation in sport. Multiple interrelated variables and feedback loops were identified, including those that create both potential positive and negative impacts. For example, multiple outcomes were identified related to the technology itself including increasing efficiency, knowledge, and generating improvements in sporting performance, and decreased time lost to injury. Despite these potential benefits, the CLD identified several negative aspects that could arise following the implementation of AI (e.g., transparency, biases, black boxism, trust), financial implications (e.g., costs of technology), a potential dependence on technology, and the contentment of humans in the system (e.g., satisfaction, job displacement, skill, and creativity degradation).

The CLD identified several reinforcing and balancing loops that will influence the implementation and use of AI in sport, including variables which are relevant to professionals currently working sport science and sports medicine teams. This illustrates the complex nature of sports science and sports medicine work, and the multiple variables and pathways which are necessary to consider when assessing the potential effects that the implementation of AI and automation technologies will have on teamwork. Lastly, from the CLD, it appears that even the most optimal application of AI in sport may not bring immediately identifiable benefits to the sports science and sports medicine multi-disciplinary teams without consideration of joint-optimisation of both the human and technological elements. A potential benefit of joint-optimisation of both human and technological elements could be the development of Multidisciplinary Human-Autonomy Teams. This could lead to similar increases in worker productivity and efficiency. It is foreseeable that, for example, medical staff would be able to query an AI regarding the symptoms that an athlete is experiencing while integrating imaging (e.g., ultrasound, MRI) data and receive a potential differential diagnosis in response (Rauscheker et al., 2020). Such human and AI teamwork integration requires new teamwork models and ways of thinking.

Overall, this case study has reinforced some of the messages from Bainbridge's (1983) seminal Ironies of Automation paper. Bainbridge highlights various unintended consequences of adopting automation, pointing to the fact that paradoxically automation increases complexity and often makes systems increasingly brittle and prone to failure. One of Bainbridge's ironies suggests that the more reliable the automation, the more reliant people become on it, potentially leading to complacency and a false sense of security. Additionally, operators of automated systems may become deskilled, unexpected situations can cause more problems, automation can reduce job satisfaction, and the complexity of automated systems can create new emergent problems. Indeed, these inherent ironies have been recast in the age of automation as the 'ironies of artificial intelligence' (Endsley, 2023). To mitigate these potential pitfalls, it's important to ensure that humans in sport systems remain involved in the operation of automated systems to prevent unintended consequences.

Recommended Reading

McLean, S., Kerhervé, H. A., Stevens, N., & Salmon, P. M. (2021). A systems analysis critique of sport-science research. *International Journal of Sports Physiology and Performance*, 16(10), 1385–1392.

Meadows, D. H. (2008). *Thinking in systems: A primer*. Chelsea Green Publishing.

Salmon, P. M., Stanton, N. A., Walker, G. H., Hulme, A., Goode, N., Thompson, J., & Read, G. J. (2022). Causal Loop Diagrams (CLDs). In *Handbook of systems thinking methods* (pp. 157–180). CRC Press.

Senge, P. M. (1990). *The fifth discipline: The art and practice of the learning organisation*. New York: Doubleday/Currency.

Sterman, J. D. (2000). *Business dynamics: Systems thinking and modeling for a complex world*. Boston: Irwin McGraw-Hill.

References

Allender, S., Owen, B., Kuhlberg, J., Lowe, J., Nagorcka-Smith, P., Whelan, J., & Bell, C. (2015). A community based systems diagram of obesity causes. *PloS One*, 10(7), e0129683.

Araújo, D., Couceiro, M., Seifert, L., Sarmento, H., & Davids, K. (2021). *Artificial intelligence in sport performance analysis*. Routledge.

Bainbridge, L. (1983). Ironies of automation. In *Analysis, design and evaluation of man–machine systems* (pp. 129–135). Elsevier: Pergamon.

Bunker, R., & Susnjak, T. (2022). The application of machine learning techniques for predicting match results in team sport: A review. *Journal of Artificial Intelligence Research*, 73, 1285–1322.
Bérard, C. (2010). Group model building using system dynamics: An analysis of methodological frameworks. *Electronic Journal of Business Research Methods*, 8(1), 35–45.
Burrell, M., White, A. M., Frerichs, L., Funchess, M., Cerulli, C., DiGiovanni, L., & Lich, K. H. (2021). Depicting 'the system': How structural racism and disenfranchisement in the United States can cause dynamics in community violence among males in urban Black communities. *Social Science & Medicine*, 272, 113469.
Chmait, N., & Westerbeek, H. (2021). Artificial intelligence and machine learning in sport research: An introduction for non-data scientists. *Frontiers in Sports and Active Living*, 3, 363.
Cust, E. E., Sweeting, A. J., Ball, K., & Robertson, S. (2019). Machine and deep learning for sport-specific movement recognition: A systematic review of model development and performance. *Journal of Sports Sciences*, 37(5), 568–600.
Endsley, M. R. (2023). Ironies of artificial intelligence. *Ergonomics*, 66(11), 1656–1668.
Hancock, P. A. (2017). Imposing limits on autonomous systems. *Ergonomics*, 60(2), 284–291.
Hancock, P. A. (2022). Avoiding adverse autonomous agent actions. *Human–Computer Interaction*, 37(3), 211–236.
Herold, M., Goes, F., Nopp, S., Bauer, P., Thompson, C., & Meyer, T. (2019). Machine learning in men's professional football: Current applications and future directions for improving attacking play. *International Journal of Sports Science & Coaching*, 14(6), 798–817.
Hovmand, P. S. (2014). *Group model building and community-based system dynamics process*. New York: Springer.
Hulme, A, Mclean S, Thompson J, Salmon PM, Lane BR, Nielsen RO. (2018). Computational methods for modelling complex systems in sports injury research: Agent-Based Modelling (ABM) and Systems Dynamics (SD) modelling. *British Journal of Sports Medicine*. DOI: 10.1136/bjsports-2018-100098.
Kim, Y. P., Seinfeld, J. H., & Saxena, P. (1993). Atmospheric gas-aerosol equilibrium I. Thermodynamic model. *Aerosol Science and Technology*, 19(2), 157–181.
Kumar, A., Priya, B., & Srivastava, S. K. (2021). Response to the COVID-19: Understanding implications of government lockdown policies. *Journal of Policy Modeling*, 43(1), 76–94.
Lin, C. L., & Chien, C. F. (2019). Systems thinking in a gas explosion accident–lessons learned from Taiwan. *Journal of Loss Prevention in the Process Industries*, 62, 103987.
Meadows, D. H. (2008). *Thinking in systems: A primer*. Chelsea Green Publishing. White River Junction, Vermont
McLean, S., Kerhervé, H. A., Stevens, N., & Salmon, P. M. (2021). A systems analysis critique of sport-science research. *International Journal of Sports Physiology and Performance*, 16(10), 1385–1392.
McLean, S., Read, G. J., Thompson, J., Baber, C., Stanton, N. A., & Salmon, P. M. (2023). The risks associated with Artificial General Intelligence: A systematic review. *Journal of Experimental & Theoretical Artificial Intelligence*, 35(5), 649–663.
Meadows, D. H. (2008). *Thinking in systems: A primer*. Chelsea Green Publishing.
Müller, V. C. (Ed.). (2016). *Risks of artificial intelligence*. CRC Press. (Vol. 5). Boca Raton, FL: CRC Press.
Naughton, M., Salmon, P. M., Compton, H. R., & McLean, S. (2024) Challenges and opportunities of artificial intelligence implementation within sports science and sports medicine teams. Frontiers in Sports and Active Living, 6, 1332427.
Omohundro, S. (2013). Rational artificial intelligence for the greater good. In *Singularity hypotheses: A scientific and philosophical assessment* (pp. 161–179). Berlin, Heidelberg: Springer.
Rauschecker, A. M., Rudie, J. D., Xie, L., Wang, J., Duong, M. T., Botzolakis, E. J., Kovalovich, A. M., Egan, J., Cook, T. C., Bryan, R. N., & Nasrallah, I. M. (2020). Artificial intelligence system approaching neuroradiologist-level differential diagnosis accuracy at brain MRI. *Radiology*, 295(3), 626–637.

Richards, C. E., Lupton, R. C., & Allwood, J. M. (2021). Re-framing the threat of global warming: An empirical causal loop diagram of climate change, food insecurity and societal collapse. *Climatic Change*, 164(3), 1–19.

Sahin, O., Salim, H., Suprun, E., Richards, R., MacAskill, S., Heilgeist, S., Rutherford, S., Stewart, R. A., & Beal, C. D. (2020). Developing a preliminary causal loop diagram for understanding the wicked complexity of the COVID-19 pandemic. *Systems*, 8(2), 20.

Salmon, P. M., Read, G. J. M., Thompson, J., McLean, S., McClure, R. (2020). Computational modelling and systems ergonomics: a system dynamics model of drink driving-related trauma prevention. *Ergonomics*, 1-47.

Salmon, P.M., McLean, S., Dallat, C., Mansfield, N., Solomon, C., Hulme, A. (2021). Human Factors and Ergonomics in Sport: Applications and Future Directions. CRC press. Boca Raton.

Salmon, P. M., Stanton, N. A., Walker, G. H., Hulme, A., Goode, N., Thompson, J., & Read, G. J. (2022). *Handbook of systems thinking methods* (pp. 157–180). CRC Press, Boca Raton, FL.

Salmon, P. M., Baber, C., Burns, C., Carden, T., Cooke, N., Cummings, M., Hancock, P., McLean, S., Read, G. J. & Stanton, N. A. (2023). Managing the risks of artificial general intelligence: A human factors and ergonomics perspective. *Human Factors and Ergonomics in Manufacturing & Service Industries*, 33(5), 366–378

Senge, P. M. (1990). *The fifth discipline: The art and practice of the learning organisation*. New York: Doubleday/Currency.

Shepherd, S. P. (2014). A review of system dynamics models applied in transportation. *Transportmetrica B: Transport Dynamics*, 2(2), 83–105.

Sterman, J. D. (2000). *Business Dynamics: Systems Thinking and Modeling for a Complex World*. Boston, MA: Irwin McGraw-Hill.

Van Eetvelde, H., Mendonça, L. D., Ley, C., Seil, R., & Tischer, T. (2021). Machine learning methods in sport injury prediction and prevention: a systematic review. *Journal of Experimental Orthopaedics*, 8, 1–15.

8

THE SYSTEMS THEORETIC ACCIDENT MODEL AND PROCESSES (STAMP) CONTROL STRUCTURE METHOD

Background

In the ever-evolving landscape of sports, a holistic understanding of underlying system structure becomes crucial for achieving optimal performance of sport systems. The world of sport is becoming more complex with technology advancements, globalisation, sports science, regulations, and policies, to name a few. Traditional linear analysis models often fall short in capturing this complexity. The Systems Theoretic Accident Model and Processes (STAMP), a method originally designed for system safety, offers a method that is suited to helping understand the complex world of sports systems.

The STAMP (Leveson, 2004) is a model of accident causation which has an associated prospective risk assessment method, the Systems Theoretical Process Analysis (STPA) (Leveson, 2011) (see Chapter 9), and retrospective accident analysis method, the Causal Analysis based on System Theory (CAST) (Leveson, 2004). According to STAMP, which is based on systems and control theory, safety or performance risks are managed through a hierarchy of controls and feedback mechanisms, and adverse events emerge when behaviour and emergent properties are not adequately controlled (Leveson, 2004). These controls are broad and include regulatory, managerial (e.g., funding), organisational (e.g., procedures), physical (e.g., engineered safety features), operational (e.g., supervision), and manufacturing-based controls (Leveson, 2004). Though the initial STAMP model is focused on safety and adverse events, it is also useful when considering system performance generally or illicit or criminal behaviour (e.g., Lane et al., 2021). When applied to sport, performance and aspects such as injury (Hulme et al., 2019), injury management (Holmes et al., 2019), and doping (McLean et al., 2023) are treated as issues of control that should be managed through a control structure that has the goal of enforcing constraints on the behaviour of key stakeholders.

Both STPA and CAST involve the development of a control structure model that depicts the control and feedback mechanisms used to manage performance in the system under analysis and the actors who share responsibility for system performance (Holmes et al., 2019). The control structure model is useful as a standalone modelling tool and is presented in this chapter independently of STPA or CAST.

DOI: 10.4324/9781003259473-10

The resulting model presents the system in the form of a hierarchy and includes all actors who share the responsibility for optimal system functioning, as well as the control and feedback mechanisms that are used to maintain optimal functioning during system design and operations. A generic control structure is presented in Figure 8.1. The control structure is based on Rasmussen's Risk Management Framework (RMF) (Rasmussen, 1997) and incorporates a series of hierarchical levels to represent the system structure. Systems are viewed as comprising interrelated components that maintain a state of dynamic equilibrium through feedback loops of control and information (Leveson, 2004). Each level therefore includes a description of the relevant actors that play a role in system design or operation. Control and feedback mechanisms are included to show what controls are enacted down the hierarchy and what information about the status of the system is sent back up the hierarchy. Within Figure 8.1 the arrows flowing down the hierarchy represent controls (Leveson, 2004) and the arrows flowing up the hierarchy represent feedback loops (Leveson, 2004).

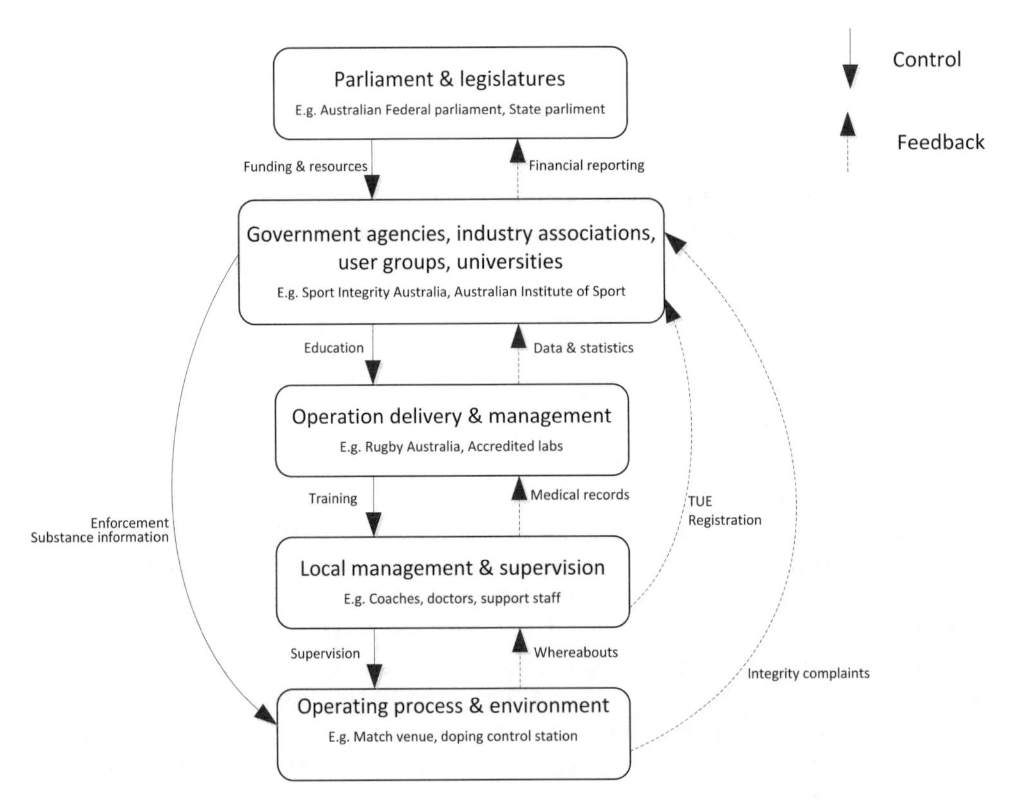

FIGURE 8.1 Truncated control structure for the anti-doping system in Australia (adapted from Leveson, 2004; McLean et al, 2023).

Applications in sport

The control structure method is a generic approach and can be applied in any domain for system modelling purposes. Beyond various safety critical domains (see Salmon et al., 2022), the method has been applied in a number of sporting contexts, including anti-doping (McLean et al., 2023), child safeguarding in sport (Dodd et al, 2024), running injury (Hulme et al., 2017), concussion in rugby (Clacy et al., 2017; Holmes et al., 2019), and talent identification in dressage (Berber et al., 2020).

Procedure and advice

A flowchart depicting the procedure for developing a control structure is presented in Figure 8.2. Step-by-step guidance is presented below.

Step 1: Define aims of the analysis and the system under analysis

The first step in developing a STAMP control structure model involves clearly defining the system under analysis, the aims of the analysis, along with any analysis boundaries. For example, the aim of the analysis may be to develop a control structure model of a particular sport system to support the analysis of a recurring issue such as injury or doping violations (e.g., Clacy et al., 2017; Holmes et al., 2019; Hulme et al., 2017; McLean et al., 2023), or it may be to develop a control structure to support analysis of a particular adverse event, or events (e.g., Dodd et al, 2024; Meng et al., 2022). Once the analysis aims are defined, the analysis boundaries should be specified. Defining the analysis boundaries is important as project constraints will dictate how deep the analysis can go in terms of the levels of the system hierarchy considered. For example, it may be that an analysis may be limited to the organisational level only (e.g., up to the level of a football club). Alternatively, the analysis may also consider government and international influences. In the anti-doping case study example presented later, the analysis boundary was set to include the sport's governing bodies, government, and international organisations.

Step 2: Data collection

The control structure model depicts:

a Which actors and organisations share the responsibility for system design and operation;
b The control mechanisms that are enacted down the system hierarchy to maintain optimal system operations; and
c The feedback mechanisms that are enacted up the system hierarchy to keep actors informed about system functioning.

To support the development of the control structure, the following data regarding system composition and system functioning is often required:

1 Data on system design and operation activities and processes;
2 Data on the work system in terms of who resides within the system and shares the responsibility for safety and optimal system functioning;
3 Data on the control and feedback mechanisms currently used to maintain optimal system functioning during system design and operation.

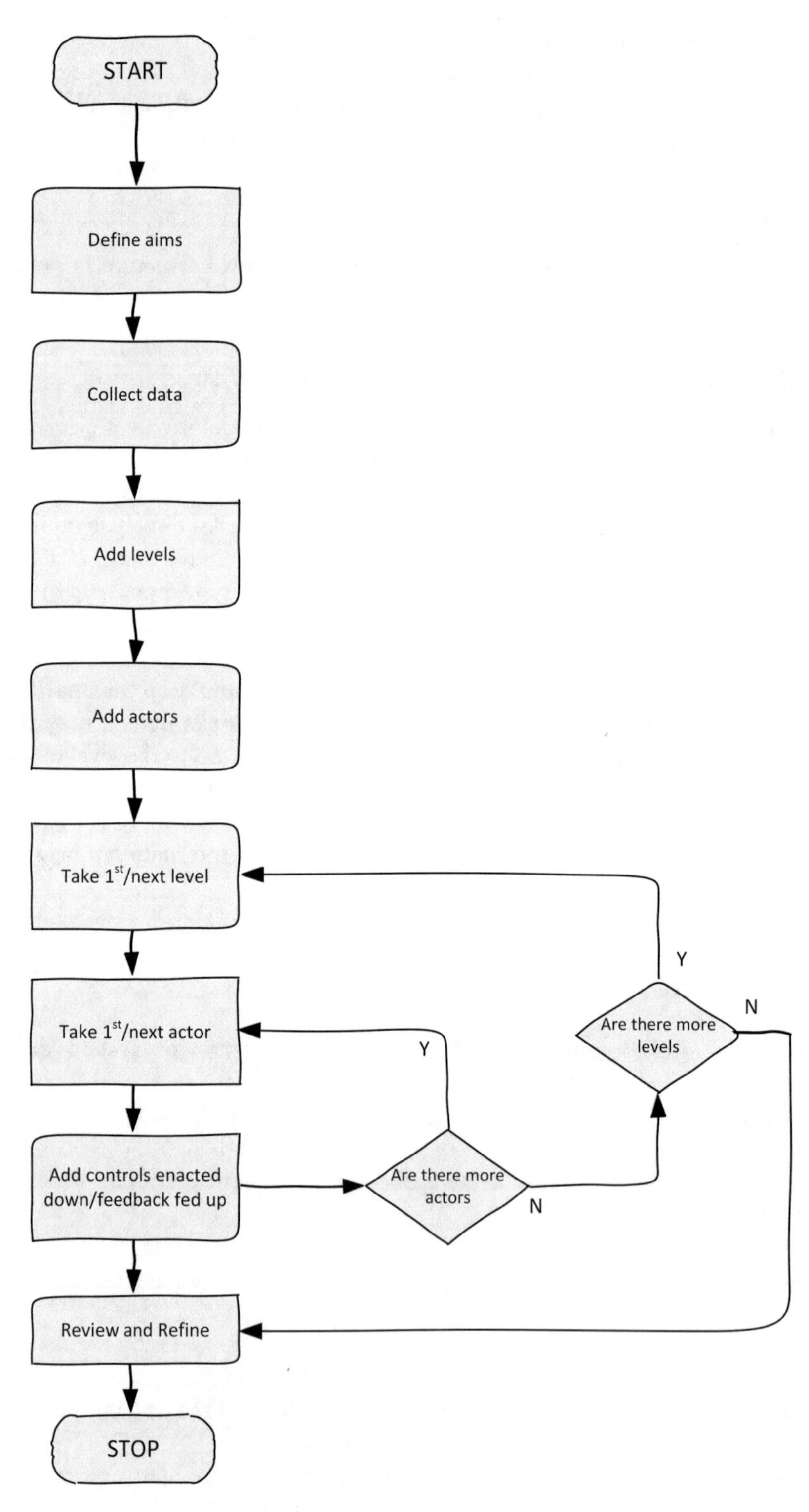

FIGURE 8.2 Control structure modelling procedure.

Data collection to support the development of a control structure model can involve a range of activities, including interviews with system actors, surveys and questionnaires, documentation review (e.g., standard operating procedures, policy and legislation, strategy documents), and observation of system design and operation activities. It is important during data collection to consider all levels of the system that are included within the analysis boundary. For example, if using interviews as the primary data collection approach it is useful to interview relevant actors from each of the levels of the system to be considered in the analysis. Likewise, when reviewing procedures this should include the procedures for different actors across all levels of the system.

Step 3: Construct draft control structure model

Once data collection is complete, a draft control structure should be developed for the system in question. This is typically undertaken using a group modelling approach involving the analyst team, or the analyst team plus an appropriate set of Subject Matter Experts (SMEs). As noted earlier, the control structure depicts the system hierarchy, relevant actors and organisations, and the control and feedback mechanisms used to maintain optimal system functioning.

Define system levels: As shown in Figure 8.1, Leveson's generic control structure model specifies a useful set of hierarchical levels for both system design and operation; however, these are flexible and can be modified depending on the aims of the analysis and the resources available. Building the draft control structure involves first specifying the system levels to be used, then identifying relevant actors and organisations at each level, and then specifying the controls and feedback mechanisms that are currently used to manage safety.

It is most useful to include both system design and system operation within the control structure; however, most analyses typically focus on the system operations control structure (see McLean et al., 2023). First, the number and name of the levels to be included in the model should be determined. The method is flexible in that any number of levels can be used; however, models are most useful when they include at least the five levels used in Leveson's generic control structure. The following levels were used in McLean et al. (2023) control structure model of the anti-doping system in four football codes in Australia:

- Level 0: International influences;
- Level 1: Parliament and legislatures;
- Level 2: Government agencies, industry associations, user groups, courts, universities;
- Level 3: Operational delivery and management;
- Level 4: local management and supervision;
- Level 5: Operating process and environment.

Define system actors: Next, relevant actors and organisations should be placed at the appropriate level in the system hierarchy. It is recommended that analysts start at the lowest level (Level 5) as this is typically the most straightforward and includes the athlete, their equipment, and the environment in which the sporting activity is undertaken.

Define system control mechanisms: Once the relevant actors and organisations have been placed at each level, control mechanisms should be added. Control mechanisms should reflect any rule, procedure, process, technology, or incentive that is used by actors at one level to control the behaviour of actors at the level below. It is worth noting that a broad view of controls should be adopted. Leveson (2004) describes various forms of control, including managerial, organisational, physical, operational, and manufacturing-based controls. In addition, external influences such as social norms and culture are often included. That is, system behaviour is controlled not only by engineered systems and direct intervention, but also by policies, procedures, shared values, and other aspects of the surrounding organisational and social culture (Leveson, 2004). As with the development of the initial system hierarchy, it is recommended that analysts start at the lowest level and work their way up the control structure. Example control mechanisms from the anti-doping in sport case study later in the chapter include the World Anti-Doping Code, targets and performance measures, anti-doping education, lab accreditation, doping control and testing, the WADA prohibited list, compliance monitoring, to name a few.

Define system feedback mechanisms: Once the control mechanisms have been finalised, feedback mechanisms going back up the hierarchy should be added. Feedback mechanisms should reflect any instances whereby information is passed from actors at one level to actors at the level above to inform them about how the system is functioning. Again, a broad view of feedback mechanisms is adopted. Examples of feedback mechanisms from the doping in sport case study include sanction data, laboratory reports, whereabouts information, therapeutic use exemption (TUE) applications, whistle-blowing, anti-doping education compliance, complaints performance data and statistics, and research findings. Again, it is recommended that analysts start at the lowest level and work their way up the control structure.

The data collected during Step 2 should be used to support development of the control structure; however, the process is iterative and often involves the identification of new data requirements. Once the draft control structure is complete it is useful to have various SMEs review it. It is best practice to use SMEs who have extensive knowledge of the system under analysis. The control structure model should be updated and finalised based on the feedback provided by the SMEs.

Step 4: Review and refine analysis

Once the control structure model is developed, it is useful to perform a formal validation exercise whereby it is reviewed and refined by appropriate SMEs. This can take on various forms, including an SME workshop (Read et al., 2018) or Delphi study (Hulme et al., 2017; Salmon et al., 2016). This involves asking SMEs to review the model and comment on the levels and level names, including system actors, controls, and feedback mechanisms. Typical refinements include modifying level and actor names, adding missing actors, controls, and feedback mechanisms, and/or removing inappropriate actors, controls, and feedback mechanisms (Dodd et al, 2024; McLean et al., 2023).

Advantages

- The control structure model output offers a detailed description of the system under analysis, including a description of system levels and actors, and the control and feedback mechanisms used to maintain optimal system functioning.
- The control structure method is based upon a robust theoretical model and is aligned with state-of-the-art control and systems theory.

- The control structure method is relatively simple to learn and use.
- The method has been applied in various sports contexts
- The control structure model can be used to support various analyses, including STAMP-CAST analyses of adverse events (e.g., serious injuries and fatalities, performance failures), prospective risk assessment via STAMP-STPA, and the analysis of control and feedback mechanisms.
- When used to support the development of interventions, the control structure model encourages analysts to identify interventions which tackle broader system issues as opposed to component fixes.

Disadvantages

- Developing and validating control structure models can be time consuming and requires access to multiple SMEs.
- The quality of the outputs is highly dependent on the quality of the data available, and the expertise of the analysts involved.
- There is little reliability and validity evidence associated with the control structure method.
- STAMP does not provide a method to support identification and development of interventions; these are based on the judgement of the analyst.
- The outputs can be complex, and large, and can be difficult to present in reports, articles and presentations.

Related methods

Various data collection methods can be used to gather the data required to develop control structure models, including interviews, observations, documentation reviews, questionnaires, and surveys. Previous studies have utilised the Delphi method, surveys, or SME workshops to validate control structure diagrams (Holmes et al., 2019; Hulme et al., 2017; McLean et al., 2023). The control structure model forms the basis for STAMP-CAST analyses of adverse events and prospective risk assessments using STAMP-STPA (see Chapter 9).

Approximate training and application times

The control structure method is simple to learn and apply, with half or full-day training workshops normally sufficient for trainees to pick up the basic concepts and approach. The method can become time-consuming when applied to complex systems, with timescales of around one to two weeks for data collection and construction of the control structure, and further time required for model validation. For example, a Delphi study can take upwards of six months to complete, as it may require multiple rounds of SME review.

Reliability and validity

Though various studies have sought to assess the reliability and validity of the STAMP-CAST and STAMP-STPA (Goncalves Filho et al., 2019; Hulme et al., 2021; 2022; 2023), to date none have attempted to assess the reliability and validity of the STAMP control structure method specifically.

Tools needed

Early drafts of the control structure model can be developed using pen paper, with the model then being drawn in a software drawing package such as Microsoft Visio.

Case study application: anti-doping in Australian Rugby Union

The use of performance-enhancing substances or 'doping' continues to represent a major issue in high-performance sports (Momaya et al., 2015; Pöppel et al., 2021). Doping occurs in a complex and dynamic environment that cannot be easily explained or understood. However, the majority of anti-doping research has focused on the athlete and much less on the complex interactions occurring across the broader sports system. As such, novel approaches to understanding the systemic influences on doping in sport may help support the development of more effective prevention efforts (McLean et al., 2023).

In this case study, we present a control structure model that was developed for the Australian Rugby Union anti-doping system. The aim was to identify the actors, controls, and feedback mechanisms currently supporting anti-doping efforts, with a view to making recommendations on how the control structure could be strengthened. The resulting control structure model was subsequently used to support a STAMP-STPA analysis of potential anti-doping failures (see Chapter 9).

Method

Eighteen people participated in the case study as anti-doping SMEs. Participants and held positions within anti-doping agencies, National Sporting Organisations, clubs and academies, and Universities. Participants held a variety of roles within anti-doping, including high-level roles of Directors, Assistant Directors, and Chief Science Officers at national and international anti-doping organisations, and specifically in anti-doping testing, football code integrity and medical programmes, sports operations, anti-doping research, sport science, medical advisors, players associations, education officers, intelligence, and TUEs.

Procedure

STAMP control structure model development

A draft control structure model of the Australian Rugby Union anti-doping system was developed by the research team based on publicly available sources, including anti-doping stakeholder websites, anti-doping policy documents, anti-doping strategies, media, and peer-reviewed literature. To accurately reflect the anti-doping system, the control structure model was adapted to include an international level, which is commonplace in other STAMP analyses (Dodd et al, 2024; Hulme et al., 2017; McLean et al, 2023; Read et al., 2019).

An on-line SME workshop was then used to review and refine the control structure model via video conferencing software with the participant group. During the workshop, participants were asked to review each of the hierarchical levels of the control structure model to determine the accuracy of the actors and organisations included in the model, and to identify any missing actors and organisations. This involved starting at the international level of the control structure and working down through the levels. Participants were then asked to review each of the controls and feedback mechanisms included in the model to determine their accuracy and to identify any missing control and feedback mechanisms. The research team then refined the model based on the feedback from the independent reviews. The refined model was sent back out to the workshop participants via email for additional comments, and further revisions were made resulting in a final validated control structure model of the Rugby Union anti-doping system in Australia.

Results

Rugby Union anti-doping control structure

The Australian Rugby Union anti-doping control structure is presented in Figure 8.3.

Discussion

The anti-doping control structure model demonstrates the complexity of the anti-doping system through the specification of numerous actors that share responsibility for anti-doping. The conceptualisation of the anti-doping system as a complex sociotechnical system supports assertions that doping activities are emergent properties of the broader sports system (Houlihan & Vidar Hanstad, 2019).

Analysis of the control structure model highlights that education for players is a key strategy for doping prevention by anti-doping agencies. However, given the diverse set of actors involved, education could be targeted not only at players but also at other system actors indirectly involved in sport and doping prevention, such as coaches, media, sponsors, supplement retailers, pharmacies, police, youth rugby participants, and recreational players. Further, numerous controls identified in the model emphasise a profoundly bureaucratic system that is focused on education, detection, deterrence, and enforcement of actors at the lower levels of the system. For example, fines, suspensions, and sanctions; education requirements; education; enforcement; training policies and procedures; surveillance; access; rules and regulations; legal penalties were among the numerous controls imposed on the lower-level actors. It may be worthwhile exploring other approaches beyond education, detection, and deterrence. For example, incentivising clean performance as well as providing anonymous and confidential mechanisms for athletes to report instances where they feel pressured to dope.

Analysis of the feedback mechanisms identified several feedbacks that are missing in the current anti-doping system. These include information regarding the effectiveness and reach of anti-doping education, benchmarking of testing frequency and prevalence statistics between different sports, and appropriate avenues for systemic feedback and reporting outside of behaviours related to anti-doping rule violations. A key recommendation is the development of an incident reporting and learning systems that look ‘up and out’ at the broader system rather than ‘down and in’ at individual wrongdoing may be required.

Overall, the findings support previous research indicating that anti-doping activities should move beyond reducing the problem of doping to a single actor, action, or piece of legislature and consider the role of the broader doping prevention system, its interrelated

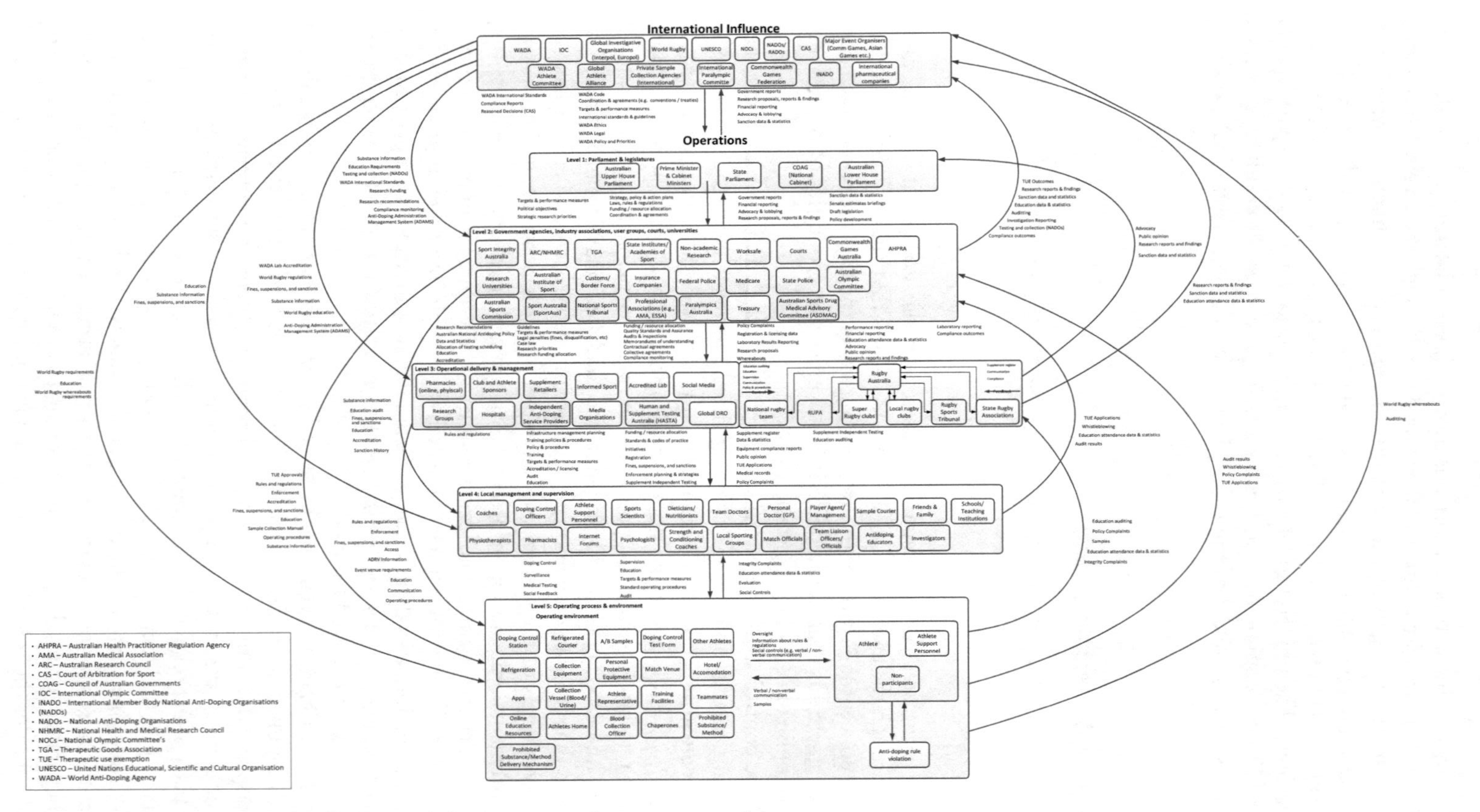

FIGURE 8.3 Australian Rugby Union anti-doping control structure model.

components, and resulting emergent properties (McLean et al, 2023; Houlihan & Vidar Hanstad, 2019). Contemporary safety science research has demonstrated that adverse incidents (e.g., doping) are caused by interacting networks of contributory factors brought about by decisions and actions from actors across the system hierarchy (Salmon et al., 2020). Systems thinking-based accident models are useful in that they are designed to help identify interacting systemic conditions that may precede and indeed create this final event (i.e., the act of doping itself). Practically, through the developed control structure model, it would be possible to identify these interacting networks of control and feedback mechanism inadequacies and the network of actors involved. Further, the implementation of systemic investigations using systems-based accident analysis would arguably impose greater and necessary accountability on the actors throughout the entire anti-doping system.

Recommended Reading

McLean, S., Naughton, M., Kerheve, H., & Salmon, P. M. (In Press). From anti-doping-I to anti-doping-II: Toward a paradigm shift for doping prevention in sport. *International Journal of Drug Policy*, 115, 104019.

Leveson, N. G. (2004). A new accident model for engineering safer systems. *Safety Science*, 42, 237–270.

Leveson, N. G. (2011). Applying systems thinking to analyze and learn from events. *Safety Science*, 49, 55–64.

Leveson, N. G. (2017). Rasmussen's legacy: A paradigm change in engineering for safety. *Applied Ergonomics*, 59, 581–591.

References

Berber, E., Beanland, V., Clacy, A., & Read, G. J. (2020). Performance pathways in the sport of dressage: A systems ergonomics approach. In Paul M. Salmon, Scott McLean, Clare Dallat, Neil Mansfield, Colin Solomon, Adam Hulme (eds) *Human Factors and Ergonomics in Sport* (pp. 269–286). CRC Press.

Clacy, A., Goode, N., Sharman, R., Lovell, G. P., & Salmon, P. M. (2017). A knock to the system: A new sociotechnical systems approach to sport-related concussion. *Journal of Sports Sciences*, 35(22), 2232–2239.

Dodd, K., Solomon, C., Naughton, M., Salmon, P. M., & McLean, S. (2024). What enables child sexual abuse in sport? A systematic review. *Trauma, Violence, & Abuse*, 25(2), 1599–1613.

Goncalves Filho, A. P., Jun, G. T., & Waterson, P. (2019). Four studies, two methods, one accident– An examination of the reliability and validity of Accimap and STAMP for accident analysis. *Safety Science*, 113, 310–317.

Holmes, G., Clacy, A., & Salmon, P. M. (2019). Sports-related concussion management as a control problem: using STAMP to examine concussion management in community rugby. *Ergonomics*, 62(11), 1485–1494.

Houlihan, B., & Vidar Hanstad, D. (2019). The effectiveness of the world anti-doping agency: Developing a framework for analysis. *International Journal of Sport Policy and Politics*, 11(2), 203–217.

Hulme, A., Stanton, N. A., Walker, G. H., Waterson, P., & Salmon, P. M. (2022). Testing the reliability and validity of risk assessment methods in human factors and ergonomics. *Ergonomics*, 65(3), 407–428.

Hulme, A., Stanton, N. A., Walker, G. H., Waterson, P., & Salmon, P. M. (2023). Testing the reliability of accident analysis methods: A comparison of AcciMap, STAMP-CAST and AcciNet. *Ergonomics*, 67(5), 695–715.

Hulme, A., Salmon, P. M., Nielsen, R. O., Read, G. J., & Finch, C. F. (2017). From control to causation: Validating a 'complex systems model' of running-related injury development and prevention. *Applied Ergonomics*, 65, 345–354.

Hulme, A., Stanton, N. A., Walker, G. H., Waterson, P., & Salmon, P. M. (2021). Are accident analysis methods fit for purpose? Testing the criterion-referenced concurrent validity of AcciMap, STAMP-CAST and AcciNet. *Safety Science*, 144.

Lane, B. R., Salmon, P. M., Desmond, D., Cherney, A., Carley, A., Hulme, A., & Stanton, N. A. (2020). Out of control? Using STAMP to model the control and feedback mechanisms surrounding identity crime in darknet marketplaces. *Applied Ergonomics*, 89, 103223.

Leveson, N. (2004). A new accident model for engineering safer systems. *Safety Science*, 42(4), 237–270.

Leveson, N. G. (2011). Applying systems thinking to analyze and learn from events. *Safety Science*, 49, 55–64.

Meng, X., Chen, G., Zhu, J., & Li, T. (2022). Application of integrated STAMP-BN in safety analysis of subsea blowout preventer. *Ocean Engineering*, 258, 111740.

McLean, S., Naughton, M., Kerheve, H., & Salmon, P. M. (2023). From anti-doping-I to anti-doping-II: Toward a paradigm shift for doping prevention in sport. *International Journal of Drug Policy*, 115, 104019.

Momaya, A., Fawal, M., & Estes, R. (2015). Performance-enhancing substances in sports: A review of the literature. *Sports Medicine*, 45(4), 517–531.

Pöppel, K., Dreiskämper, D., & Strauss, B. (2021). Breaking bad: How crisis communication, dissemination channel and prevalence influence the public perception of doping cases. *Sport in Society*, 24(7), 1156–1182.

Rasmussen, J. (1997). Risk management in a dynamic society: a modelling problem. *Safety Science*, 27(2), 183–213.

Read, G. J., Naweed, A., & Salmon, P. M. (2019). Complexity on the rails: A systems-based approach to understanding safety management in rail transport. *Reliability Engineering System Safety*, 188, 352–365.

Read, G. J., Salmon, P. M., Goode, N., & Lenné, M. G. (2018). A sociotechnical design toolkit for bridging the gap between systems-based analyses and system design. *Human Factors Ergonomics in Manufacturing Service Industries*, 28(6), 327–341.

Salmon, P. M., Hulme, A., Walker, G. H., Waterson, P., Berber, E., & Stanton, N. A. (2020). The big picture on accident causation: A review, synthesis and meta-analysis of AcciMap studies. *Safety Science*, 126, 104650.

Salmon, P. M., Read, G. J., & Stevens, N. J. (2016). Who is in control of road safety? A STAMP control structure analysis of the road transport system in Queensland, Australia. *Accident Analysis & Prevention*, 96, 140–151.

PART 3

Systemic Risk and Accident Analysis Methods

9

THE SYSTEMS THEORETIC ACCIDENT MODEL AND PROCESSES (STAMP)-SYSTEMS THEORETIC PROCESS ANALYSIS (STPA) METHOD

Background

In the complex world of professional sports, the interplay of athletes, coaches, organisational dynamics, and technological systems can create a multitude of physical and health risks, performance and safety risks, financial, reputational, and operational risks, as well as a raft of evolving technological risks. Prospective risk assessment in sport is thus not only about understanding performance risks, but also about understanding the broad spectrum of risks that could emerge across entire sports systems. Comprehensive methods for risk identification are therefore not just valuable, they are vital for sports clubs, associations, and governing bodies. The Systems Theoretic Process Analysis (STPA) (Leveson, 2011), is one such method that offers a systematic approach to identifying the risks inherent within modern sports systems.

The STPA is a prospective risk assessment method derived from the System Theoretic Accident Model and Processes (STAMP) Leveson, 2004) (Chapter 8). The STAMP is a model of accident causation which has an associated prospective risk assessment method – STPA (Leveson, 2011). According to STAMP, which is based on systems and control theory, risks are managed through a hierarchy of controls and feedback mechanisms, and adverse events emerge when behaviour and emergent properties are not adequately controlled (Leveson, 2004). These controls are broad and include regulatory, managerial (e.g., funding), organisational (e.g., procedures), physical (e.g., engineered safety features), operational (e.g., supervision), and manufacturing-based controls (Leveson, 2004). Though the initial STAMP model is focused on safety and adverse events, it is also useful when considering system performance generally or illicit or criminal behaviour (Lane et al., 2020). When applied to sport, performance, and aspects such as injury (Hulme et al., 2019), injury management (Holmes et al., 2019), and doping (McLean et al., 2023) are treated as issues of control that should be managed through a control structure that has the goal of enforcing constraints on the behaviour of key stakeholders.

DOI: 10.4324/9781003259473-12

STAMP-STPA is used to forecast instances where control and feedback mechanisms that are inadequate or could potentially fail, with the outputs supporting the development of risk controls designed to prevent or manage identified failures. Applying STAMP-STPA involves developing a control structure model of the system in question and then applying a control and feedback failure taxonomy to identify potential control and feedback failures. The control structure model (see Chapter 8) depicts the control and feedback relationships that exist between actors and organisations during system design and operation. A generic control structure is presented in Figure 9.1. The control structure is based on Rasmussen's risk management framework (Rasmussen, 1997) and incorporates a series of hierarchical levels to represent the system structure. Systems are viewed as comprising interrelated components that maintain a state of dynamic equilibrium through feedback loops of control and information (Leveson, 2004). Each level therefore includes a description of the relevant actors that play a role in system design or operation. Control and feedback mechanisms are included to show what controls are enacted down the hierarchy and what information about the status of the system is sent back up the hierarchy. Within Figure 9.1 the arrows flowing down the hierarchy represent controls, and the arrows flowing up the hierarchy represent feedback mechanisms (Leveson, 2004).

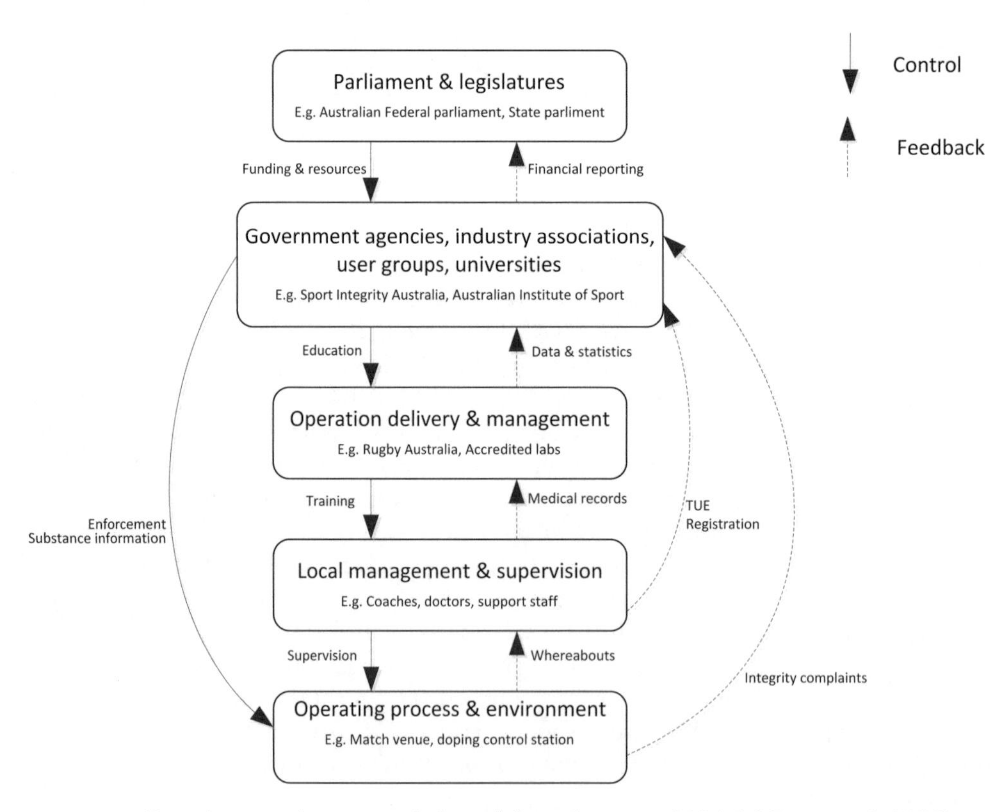

FIGURE 9.1 Generic control structure (adapted from Leveson, 2004; McLean et al, 2023).

STAMP-STPA works by considering each of the control and feedback mechanisms from the control structure along with a control/feedback failure taxonomy comprising the following failure modes (Leveson, 2011):

1. Control or feedback action is not provided or followed e.g. an injury prevention procedure does not exist, or is not followed;
2. Unsafe control or feedback action is provided, e.g., an injury rehabilitation procedure encourages potential injurious behaviour;
3. Control or feedback action is provided too early or too late, e.g., an injury prevention procedure is adopted before it is empirically tested, or is adopted too late after it is deemed appropriate;
4. Control or feedback action is stopped too soon, or applied too long, e.g., doping education is provided too late resulting in stakeholders who are relying on out-of-date education information, or education applied too long resulting in outdated approaches being used.

Where credible risks are identified, the analyst provides a description of the failure, its consequences on system functioning, and provides recommendations regarding appropriate risk controls.

Domain of application

STAMP-STPA is a generic approach and can be applied in any domain for prospective risk assessment purposes. To date STPA has been applied in a diverse set of domains, including elite women's cycling (Hulme et al., 2021a), cybersecurity (Lane et al., 2020), maritime (Chaal et al., 2020), healthcare (Bas, 2020), rail (Yang et al., 2019) aviation (Castilho et al., 2018), and space operations (Rising & Leveson, 2018).

Procedure and advice

A flowchart depicting the STAMP-STPA procedure is presented in Figure 9.2. Step-by-step guidance is presented below.

Step 1: Define aims of the analysis and the system under analysis

The first step in applying STAMP-STPA involves clearly defining the aims of the analysis along with the system and any analysis boundaries. For example, the aim of the analysis may be to identify new control and feedback mechanisms in a particular context, or alternatively it may be to develop a comprehensive risk register for a particular system. Once the analysis aims are defined, the analysis boundaries should be specified. Defining the analysis boundaries is important as project constraints will dictate how detailed the analysis can go in terms of the components of the system considered. It may be, for example, that an analysis may be limited to the organisational level only (e.g., sports club performance). In the case study example presented later, the analysis boundary was set to include sports governing bodies, government, and international organisation.

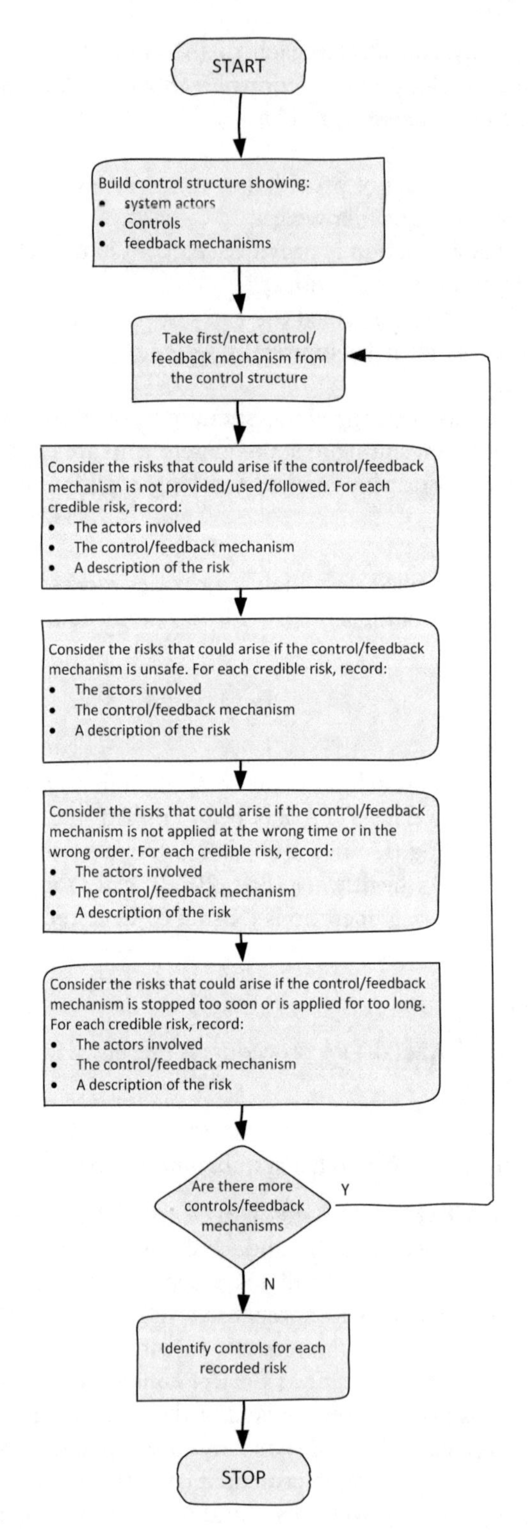

FIGURE 9.2 STAMP-STPA procedure.

Step 2: Data collection

The STAMP-STPA analysis is dependent upon the development of an accurate control structure which depicts:

a Which actors and organisations share the responsibility for optimal system design and operation;
b The control mechanisms that are enacted down the hierarchy to maintain optimal system operations; and
c The feedback mechanisms that are enacted up the hierarchy to keep actors informed about system functioning.

To support development of the control structure, data regarding the system and system functioning is required. For STAMP-STPA, two forms of data are required:

1 Data on system design and operation activities or processes;
2 Data on the work system in terms of who resides within the system and shares the responsibility for safety and optimal system functioning;
3 Data on the control and feedback mechanisms currently used to maintain optimal system functioning during system design and operation.

Data collection for STAMP-STPA can involve a range of activities, including interviews with system actors, surveys and questionnaires, documentation review (e.g., Standard operating procedures, policy and legislation, strategy documents), and observation of system design and operation. It is important during data collection to consider all levels of the system, up to and including regulatory, government and even international levels and influences. For example, if using interviews as the primary data collection approach it is useful to interview relevant actors from all levels of the system. Likewise, when reviewing procedures this should include the procedures for different actors across all levels of the system.

Step 3: Construct and verify control structure

A full procedure for the development control structure models is presented in Chapter 8. Once data collection is complete, a draft control structure should be developed for the system in question. As noted earlier, the control structure depicts the system hierarchy, relevant actors and organisations, and the control and feedback mechanisms used to maintain optimal system functioning (see Figure 9.1 for generic control structure). As shown in Figure 9.1, the generic control structure specifies a set of hierarchical levels, however, these are flexible and can be modified depending on the aims of the analysis and the resources available. Building the control structure involves first specifying the system levels, then identifying relevant actors and organisations at each level, and then specifying the controls and feedback mechanisms that are currently used to manage safety. Where possible, it is recommended that a group modelling process is used involving multiple analysts and/or SMEs. The data collected during Step 2 should be used to support the development of the control structure; however, the process is iterative and often involves the identification of new data requirements. Once developed, it is useful to subject the control structure to a validation process whereby it is reviewed and refined by appropriate SMEs. This can take on various forms, including an SME workshop (Read et al., 2018) or Delphi study (Salmon et al., 2016).

Step 4: Identify potential control failures

The first analysis step involves identifying potential control failures by considering the following four potential failure modes for each of the control mechanisms in the control structure (Leveson, 2011):

1 Control action is not provided or followed.
2 Inappropriate control action is provided.
3 Control action is provided too early or too late.
4 Control action is stopped too soon or applied too long.

During this step the analyst takes each control mechanism and considers whether there are credible risks associated with each of the four failure modes. For example, when applying STAMP-STPA to the rugby anti-doping control structure presented later, when examining the control mechanism 'Drug testing' the analyst would consider whether any of the four control failure modes are credible and would create a risk to optimal anti-doping system functioning. Credible risks for the drug testing control mechanism could include:

1 Athletes are not drug tested which could enable doping violations to occur and not be detected.
2 The drug testing procedure is inappropriate and could potentially damage athlete health and wellbeing.
3 The testing procedure is delayed and is performed too late and athlete doping violations can potentially go undetected.
4 Drug testing is stopped during out of competition periods, enabling athletes to commit doping violations without detection.

For each credible control failure, the analyst should provide a description in the STAMP-STPA output table (see Table 9.1). The output table includes a description of potential control failures and their consequences which is used to support the identification, development, and implementation of appropriate controls in Step 6.

TABLE 9.1 STAMP-STPA output table for control mechanisms

From	*To*	*Control*	*Action required but not provided*	*Unsafe action provided*	*Incorrect/ timing order*	*Stopped too soon/applied too long*

Step 5: Identify potential feedback failures

Once all of the controls within the control structure have been assessed, the STPA process should be repeated for all of the feedback mechanisms included within the control structure. This involves the analyst examining each feedback mechanism and considering whether there are credible risks associated with each of the four failure modes. For example, for the feedback mechanism 'therapeutic use exemption (TUE) application' the analyst would consider whether any of the four failure modes are credible and could create a risk to optimal anti-doping system functioning. Credible risks for the TUE feedback mechanism could include:

1 Application for TUE is not made, resulting in potential anti-doping violation.
2 TUE application does not include all information, potentially resulting in inappropriate granting of the TUE.
3 TUE application is delayed and is submitted after the fact, resulting in a potential doping violation.
4 TUE is granted for too long, which could allow overuse and the means to commit a doping violation.

As with step 4, for each credible risk, the analyst should provide a description in the STPA output table (see Table 9.2).

Step 6: Review and refine analysis

Once the draft STAMP-STPA analysis is complete it is useful to have various SMEs review it. It is best practice to use SMEs who have extensive knowledge of the system and activities in question. The STAMP-STPA analysis should be updated and finalised based on the feedback provided by the SMEs. It can also be useful for reliability purposes to have a second analyst identify control and feedback failures for a sub-set of the control structure. These can then be compared to the initial analysis to determine analysis reliability.

Step 7: Identify risk controls

Once the analyst is satisfied that all credible risks associated with control and feedback failures have been identified, the STAMP-STPA output should be used to support the identification of recommended risk controls. This typically involves a workshop or group session where each risk is considered, and potential controls are discussed. It is also

TABLE 9.2 STAMP-STPA output table for feedback mechanisms

From	*To*	*Feedback*	*Action required but not provided*	*Unsafe action provided*	*Incorrect/ timing order*	*Stopped too soon/applied too long*

useful at this stage to develop an action plan which includes a description of each control, the criteria for evaluating the success of each control, responsibilities for implementing and monitoring each control, and a review timeline.

Advantages

- The STAMP control structure model provides a useful and comprehensive description of the system, including relevant actors and organisations and control and feedback mechanisms;
- STAMP-STPA is based upon a sound theoretical model and is aligned with state-of-the-art risk and accident causation theory;
- STAMP-STPA enables the prospective identification of control and feedback failures for any given system;
- The consideration of feedback failures and their associated risks is novel and is a key strength over other risk assessment methods;
- Reliability and validity evidence is sound (e.g., Hulme et al., 2021; 2022);
- The use of a control and feedback failure taxonomy potentially enhances reliability and validity;
- The output offers a detailed analysis of potential risks within the system under analysis;
- STAMP-STPA is a generic approach which has been applied across many domains, including elite sport (e.g., Hulme et al., 2021).

Disadvantages

- STAMP-STPA can be time-consuming to apply, involving the development of a system control structure followed by the identification of potential control and feedback failures;
- The analysis does not explicitly consider emergent risks that arise when control and feedback failures interact with one another;
- The quality of the outputs is highly dependent on the expertise of the analyst(s) and the validity of the control structure model;
- STAMP-STPA does not provide a formal or structured method to support identification and development of risk controls; and
- The outputs can be complex and large, with hundreds of potential control and feedback failures often identified.

Related methods

Various data collection methods can be used to gather the data required for STAMP-STPA analyses, including interviews, observation, documentation review, questionnaire, and surveys (see Stanton et al., 2013). The STAMP control structure method is useful in its own right as a complex systems modelling method (Read et al., 2018; Salmon et al., 2016). Previous applications have used the Delphi method or SME workshops to validate control structure diagrams (Read et al., 2018; Salmon et al., 2016).

Approximate training and application times

STAMP-STPA is relatively simple to learn and apply but can become time-consuming when applied to complex scenarios or full sport systems. Development of the control structure can be time-consuming, with timescales of around one to two weeks for data collection and construction of the control structure, and further time required for model validation. Identification of control and feedback failures can also be time-consuming and is dependent on the number of control and feedback mechanisms identified.

Reliability and validity

Hulme et al. (2021b) tested the criterion-referenced validity of STAMP-STPA, EAST BL (Chapter 5), and Net-HARMS (Chapter 10). This involved training novice participants in one of the three methods and then asking them to use the method to identify the risks across a railway-level crossing design lifecycle. Participants then returned four weeks later to complete the same training and analysis. Hulme et al. (2022) found moderate levels of validity for STAMP-STPA, suggesting that participants were able to achieve moderate levels of agreement with an expert 'gold standard' analysis.

Tools needed

STPA can be conducted using pen and paper; however, the control structure is normally created in a software drawing package such as Microsoft Visio and the STPA output table is typically created in Microsoft Word or Excel.

Case study application: anti-doping in Australian Rugby Union

Background

In this case study we present a STAMP-STPA analysis for the Australian Rugby Union anti-doping system. The use of performance enhancing substances or 'doping' continues to represent a major risk in high-performance sports. Doping occurs in a complex and dynamic environment that comprises multiple hierarchical levels of the system that cannot be easily explained or understood (McLean et al, 2023). However, the majority of research on doping has focused on the athlete and much less on the complex set of interactions occurring across the broader sports system. As such, novel approaches to understanding the systemic influences on doping in sport may help support the development of more effective prevention efforts. Further, no systemic risk assessments have been applied to doping in sport.

The aim of the present case study was to apply the STPA process to a STAMP control structure model of the anti-doping system in Australian Rugby Union (see Chapter 8) to identify potential control and feedback failures, with a view to identifying opportunities for new controls and feedback mechanisms.

Method

STAMP model development

A draft control structure model of the Australian Rugby Union anti-doping system was developed by the research team based on publicly available sources, including anti-doping stakeholder websites, anti-doping policy documents, anti-doping strategies, media, and peer-reviewed literature. To accurately reflect the anti-doping system, the control structure model was adapted to include an international level, which is commonplace in other STAMP analyses (Hulme et al., 2017; Read et al., 2019). See Chapter 8 for specific details on the STAMP control structure development.

STAMP-STPA

One author conducted the STPA, which was reviewed by two other authors. The process involved taking each control and feedback from the control structure model and considering any credible risks associated with the four failure modes. For example, for the control 'world anti-doping code (WADC)' which is enacted from the International level on the Government Agencies level, potential control failures are included (Table 9.3). This process was followed for each control mechanism throughout the control structure model.

Once all of the controls within the control structure had been assessed, the process was repeated for each of the feedback mechanisms included in the model. For example, potential feedback failures for 'Compliance outcomes' which are passed from the Government Agencies level to the International influences level include (Table 9.4).

This process was followed for each feedback mechanism throughout the control structure model.

Results

Rugby Union anti-doping control structure

The Australian Rugby Union anti-doping control structure is presented in Figure 9.3.

STAMP-STPA

A total of 712 risks were identified, including 428 control risks and 284 feedback risks. Extracts of the control and feedback failures are presented in Tables 9.5 and 9.6.

TABLE 9.3 Example STAMP-STPA output table for the control mechanism 'World anti-doping code'

From	*To*	*Control*	*Action required but not provided*	*Unsafe action provided*	*Incorrect/timing order*	*Stopped too soon/applied too long*
International level	Government agencies	WADC	No unified approach to anti-doping is provided resulting in a lack of guidance to countries, and a weaker anti-doping system.	Inappropriate or inadequate WADC is provided resulting in sub-standard anti-doping policy, rules & regulations for anti-doping.	WADC is implemented or disseminated too late meaning that it is not enacted or followed for a period of time.	The WADC is applied for too long without modification and thus does not consider emerging issues such as new forms of performance-enhancing drugs, new detection prevention activities, and policy changes elsewhere in the system.

TABLE 9.4 Example STAMP-STPA output table for the control mechanism 'World anti-doping code'

From	*To*	*feedback*	*Action required but not provided*	*Unsafe action provided*	*Incorrect/timing order*	*Stopped too soon/applied too long*
Government Agencies	International influences	Compliance outcomes	No compliance outcomes are provided so stakeholders unaware of if they are compliant with current guidance.	Inaccurate compliance outcomes are provided leading to a lack of clarity as to the accurate state of stakeholder's compliance with guidance.	Compliance outcomes are provided too late, and compliance deteriorates as a result of less monitoring.	Compliance outcome recordings are stopped too soon leading to inability to determine current compliance outcomes and inaccurate guidance resulting.

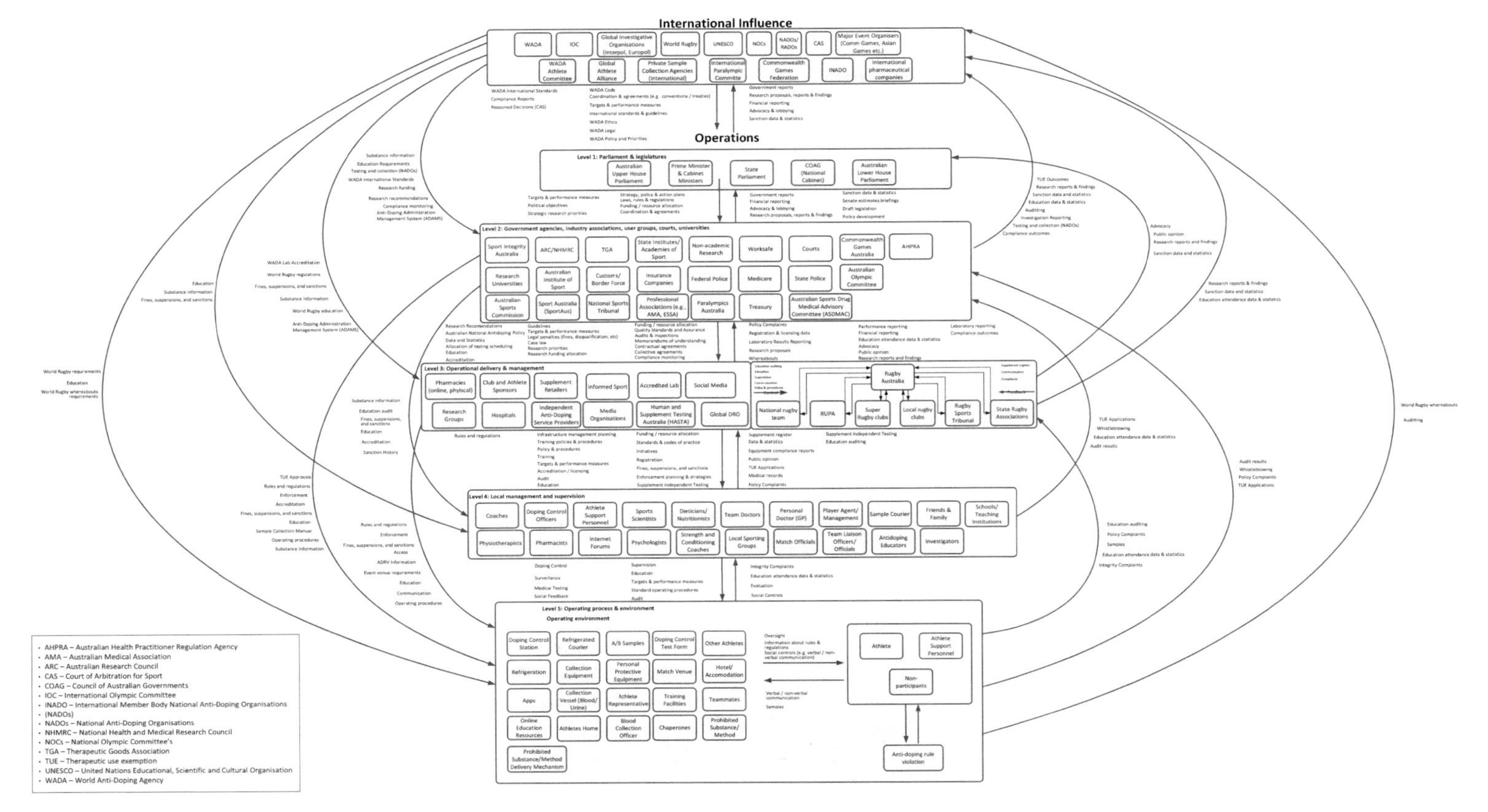

FIGURE 9.3 Australian Rugby Union anti-doping control structure.

TABLE 9.5 Example control failures taken from various levels of the STAMP control structure model

From	*To*	*Control*	*Action required but not provided*	*Unsafe action provided*	*Incorrect/timing order*	*Stopped too soon/ applied too long*
International influence	Parliament and legislatures	Compliance Reports	No compliance reports provided resulting in inability to determine if countries are compliant with WADC resulting in deterioration of the anti-doping system.	Inappropriate compliance reports provided meaning incorrect international guidance for anti-doping.	Compliance reports provided too late resulting in delayed reporting of compliance activities to inform KPIs, risks & challenges.	Compliance reports are provided too early prior to identifying new risks and challenges.
Parliament and legislatures	Government agencies	Funding/ resource allocation	Funding and resources not allocated anti-doping practices and detection methods aren't implemented.	Funding/resource allocation not adequate to address sufficiently address ADVRs.	Funding provided too late resulting in a lack of funding for testing to address ADRVs	Funding and resources allocation stopped too soon resulting in the cessation of work.
Government agencies	Operational delivery & management	Allocation of testing scheduling	No testing schedule allocated meaning no testing takes place.	Inappropriate testing schedule allocated provided meaning testing is insufficient.	Testing schedule being provided too late resulting in athletes not being tested and testing being insufficient.	Testing schedule allocation being stopped too soon resulting in athletes not being tested and testing being insufficient.
Local management & supervision	Operating process & environment	Education	No education provided resulting in ineffective education provided and an increase in athlete doping.	Inappropriate education provided resulting in uneducated stakeholders.	Education being provided too late resulting in stakeholders who are relying on out-of-date education information.	Education applied too long resulting in outdated approaches being used.

TABLE 9.6 Example feedback failures taken from various levels of the STAMP control structure model

From	*To*	*Control*	*Action required but not provided*	*Unsafe action provided*	*Incorrect/timing order*	*Stopped too soon/ applied too long*
Operating process & environment	Local management & supervision	Education attendance data	No education attendance data provided resulting in an inability to determine who is attending education and compliant with current education requirements.	Inaccurate education attendance data provided leading to an inaccurate understanding of who has been attending education.	Education attendance data provided too late to accurately capture stakeholders who attended.	Education attendance data recording stopped too soon, and attendees are not recorded leading to them being viewed as uneducated.
Local management & supervision	Operational delivery & management	Supplement Independent Testing	No supplement independent testing provided resulting in potentially tainted supplements being given and consumed.	Inaccurate Supplement Independent Testing resulting in potentially tainted supplements being approved and consumed.	Supplement independent testing is provided too late and potentially tainted supplements are consumed before testing and approval.	Supplement Independent Testing is stopped too soon, and supplements are not tested leading to inadvertent doping.
Operational delivery & management	Government agencies	Compliance Outcomes	No compliance outcomes provided so stakeholders unaware of if they are compliant with current guidance.	Inappropriate compliance outcomes communicated leading to confusion about whether stakeholders are compliant with up-to-date guidance.	Compliance outcomes are provided too late, and compliance deteriorates as a result of less monitoring.	Compliance outcomes recordings stopped too soon leading to inability to determine current compliance outcomes and inaccurate guidance resulting.
Parliament & legislatures	International influence	Sanction data & statistics	No sanction data & statistics provided leading to an inability to track trends in testing and doping rates.	Inaccurate sanction data & statistic leading to the making of guidance that does not improve current conditions and outcomes for stakeholders.	Sanction data & statistics being provided too late resulting in policies which do not address the accurate state of current sanctions	Sanction data & statistics being applied too long resulting in sanctions being applied for athletes whose sanctions should have finished.

Discussion

The STAMP-STPA analysis provides useful insights about the different ways in which the anti-doping system does fail and could potentially fail. The STPA analysis also provides risk descriptions that are relevant to numerous stakeholders within the anti-doping system, identified in the STAMP control structure model. This provides further evidence regarding the notion that there is a shared responsibility for doping that spans multiple stakeholders across sport systems.

The STPA revealed a total of 712 potential failures, including 428 control failures and 284 feedback failures. While some of these failures may be currently well managed through strong control and feedback mechanisms, some may not be. Consequently, the list of identified potential failures provides a useful database for review when considering the effectiveness of current control and feedback mechanisms and the development of new interventions. It is recommended that sport system stakeholders review the STPA risks and their current control and feedback mechanisms to determine if further effort around the development or strengthening of control and feedback mechanisms is required.

For the controls, while almost a third of the potential failures relate to the operating process and environment level and the controls imposed on athletes (e.g., testing, education, enforcement), the majority relate to controls imposed on other actors in the anti-doping sport system. For example, 76 potential control failures were identified for the controls imposed by actors at the government agencies level on actors at the operational delivery and management level. These included potential failures around the quality and implementation of the national anti-doping policy, compliance monitoring, resource allocation, quality standards and assurance, and accreditation. Similarly, 40 potential control failures were identified for the controls imposed by international actors on actors at the parliament and legislatures level. For example, these included potential failures relating to the content and implementation of the WADC, ethics guidelines, and policies and procedures. These findings demonstrate the critical need to consider risk assessment and management for all stakeholders within the anti-doping system as opposed to athlete only controls. While testing currently provides a strong control for athlete doping, there may be other opportunities to strengthen the anti-doping system. For example, enhancing controls such as 'funding and resource allocation' and 'allocation of testing schedule' will in turn enhance testing outcomes, as more athletes can be tested in an appropriate manner, and testing protocols can keep pace with advances in performance-enhancing drugs.

Recommended Reading

Leveson, N. G., & Thomas, J. P. (2018). *STPA handbook*. Cambridge, MA: MIT.

Leveson, N. J. (2004). A new accident model for engineering safer systems. *Safety Science*, 42(4), 237–270.

References

Bas, E. (2020). STPA methodology in a socio-technical system of monitoring and tracking diabetes mellitus. *Applied Ergonomics*, 89, 103190.

Canham, A., Jun, G. T., Waterson, P., & Khalid, S. (2018). Integrating systemic accident analysis into patient safety incident investigation practices. *Applied Ergonomics*, 72, 1–9.

Castilho, D. S., Urbina, L. M., & de Andrade, D. (2018). STPA for continuous controls: A flight testing study of aircraft crosswind takeoffs. *Safety Science*, 108, 129–139.
Chaal, M., Banda, O. A. V., Glomsrud, J. A., Basnet, S., Hirdaris, S., & Kujala, P. (2020). A framework to model the STPA hierarchical control structure of an autonomous ship. *Safety Science*, 132, 104939.
Holmes, G., Clacy, A., & Salmon, P. M. (2019). Sports-related concussion management as a control problem: using STAMP to examine concussion management in community rugby. *Ergonomics*, 62(11), 1485–1494.
Hulme, A., McLean, S., Dallat, C., Walker, G. H., Waterson, P., Stanton, N. A., & Salmon, P. M. (2021a). Systems thinking-based risk assessment methods applied to sports performance: A comparison of STPA, EAST-BL, and Net-HARMS in the context of elite women's road cycling. *Applied Ergonomics*, 91, 103297.
Hulme, A., Salmon, P. M., Nielsen, R. O., Read, G. J. M., & Finch, C. F. (2017). From control to causation: Validating a 'complex systems model' of running-related injury development and prevention. *Applied Ergonomics*, 65, 345–354.
Hulme, A., Stanton, N. A., Walker, G. H., Waterson, P., & Salmon, P. M. (2021b). Testing the reliability and validity of risk assessment methods in. *Human Factors and Ergonomics*, 65(3), 407–428.
Lane, B. R., Salmon, P. M., Desmond, D., Cherney, A., Carley, A., Hulme, A., & Stanton, N. A. (2020). Out of control? Using STAMP to model the control and feedback mechanisms surrounding identity crime in darknet marketplaces. *Applied Ergonomics*, 89, 103223.
Leveson, N. G. (2011). *Engineering a safer world: Systems thinking applied to safety*. Cambridge, MA: MIT.
Leveson, N. G., & Thomas, J. P. (2018). *STPA handbook*. Cambridge, MA: MIT.
Leveson, N. J. (2004). A new accident model for engineering safer systems. *Safety Science*, 42(4), 237–270.
McLean, S., Naughton, M., Kerheve, H., & Salmon, P. M. (2023). From anti-doping-I to anti-doping-II: Toward a paradigm shift for doping prevention in sport. *International Journal of Drug Policy*, 115, 104019.
Rasmussen, J. (1997). Risk management in a dynamic society: A modelling problem. *Safety Science*, 27(2–3): 183–213.
Read, G. J., Naweed, A., & Salmon, P. M. (2019). Complexity on the rails: A systems-based approach to understanding safety management in rail transport. *Reliability Engineering System Safety*, 188, 352–365.
Read, G. J., Salmon, P. M., Goode, N., & Lenné, M. G. (2018). A sociotechnical design toolkit for bridging the gap between systems-based analyses and system design. *Human Factors Ergonomics in Manufacturing Service Industries*, 28(6), 327–341.
Rising, J. M., & Leveson, N. G. (2018). Systems-theoretic process analysis of space launch vehicles. *Journal of Space Safety Engineering*, 5(3–4), 153–183.
Salmon, P. M., Read, G. J. M., & Stevens, N. J. (2016). Who is in control of road safety? A STAMP control structure analysis of the road transport system in Queensland, Australia. *Accident Analysis and Prevention*, 96, 140–151.
Yang, P., Karashima, R., Okano, K., & Ogata, S. (2019). Automated inspection method for an STAMP/STPA-fallen barrier trap at railroad crossing. *Procedia Computer Science*, 159, 1165–1174.

10

THE NETWORKED HAZARD ANALYSIS AND RISK MANAGEMENT SYSTEM (Net-HARMS)

Background

Interactions between athletes, coaches, performance and administrative staff, executives, stakeholders, and fans within a sporting club create a dynamic entity that produces emergent behaviours (Salmon & McLean, 2020). As highlighted throughout this book, this emergence is derived not simply from the sum of individual parts but from their dynamic interactions. Where these interactions are not optimal or not managed appropriately, the resulting emergent behaviours can be hard to predict. This unpredictability makes it challenging to identify, understand, and manage emergent risks.

The Networked Hazard Analysis and Risk Management System (Net-HARMS) (Dallat et al., 2018) is a prospective risk assessment method that was developed to assist organisations in identifying potential risks and developing effective controls. The method combines hierarchical task analysis (HTA) (Annett et al., 1971) (Chapter 3) with principles of the event analysis of systemic teamwork (EAST) framework (Stanton et al., 2018) (Chapter 5) and the systematic human error reduction and prediction approach (SHERPA) (Embrey, 1986). Applying Net-HARMS involves first developing a network-based model of the interrelated tasks undertaken within the system, followed by the application of a risk mode taxonomy to identify task risks and emergent risks (Salmon et al., 2022).

The Net-HARMS method is considered state-of-the-art as it enables the comprehensive identification of risks across entire sociotechnical systems and also supports the identification of 'emergent' risks – risks that emerge when other system risks interact with one another. In a sports context, the ability to identify risks across a system is useful as analysts can use Net-HARMS to identify risks not only relating to athlete and coach behaviour/activities, but also relating to other stakeholders such as club management, executives, sponsors, and governing bodies. Further, the ability to identify emergent risks is a critical requirement for understanding and managing risk in complex systems. Such risks are difficult to identify and are often not dealt with by standard risk controls.

Net-HARMS was designed with the intention of creating a method for practitioners that is easy to learn and use (Dallat et al., 2018), see Net-HARMS procedure (Figure 10.1).

DOI: 10.4324/9781003259473-13

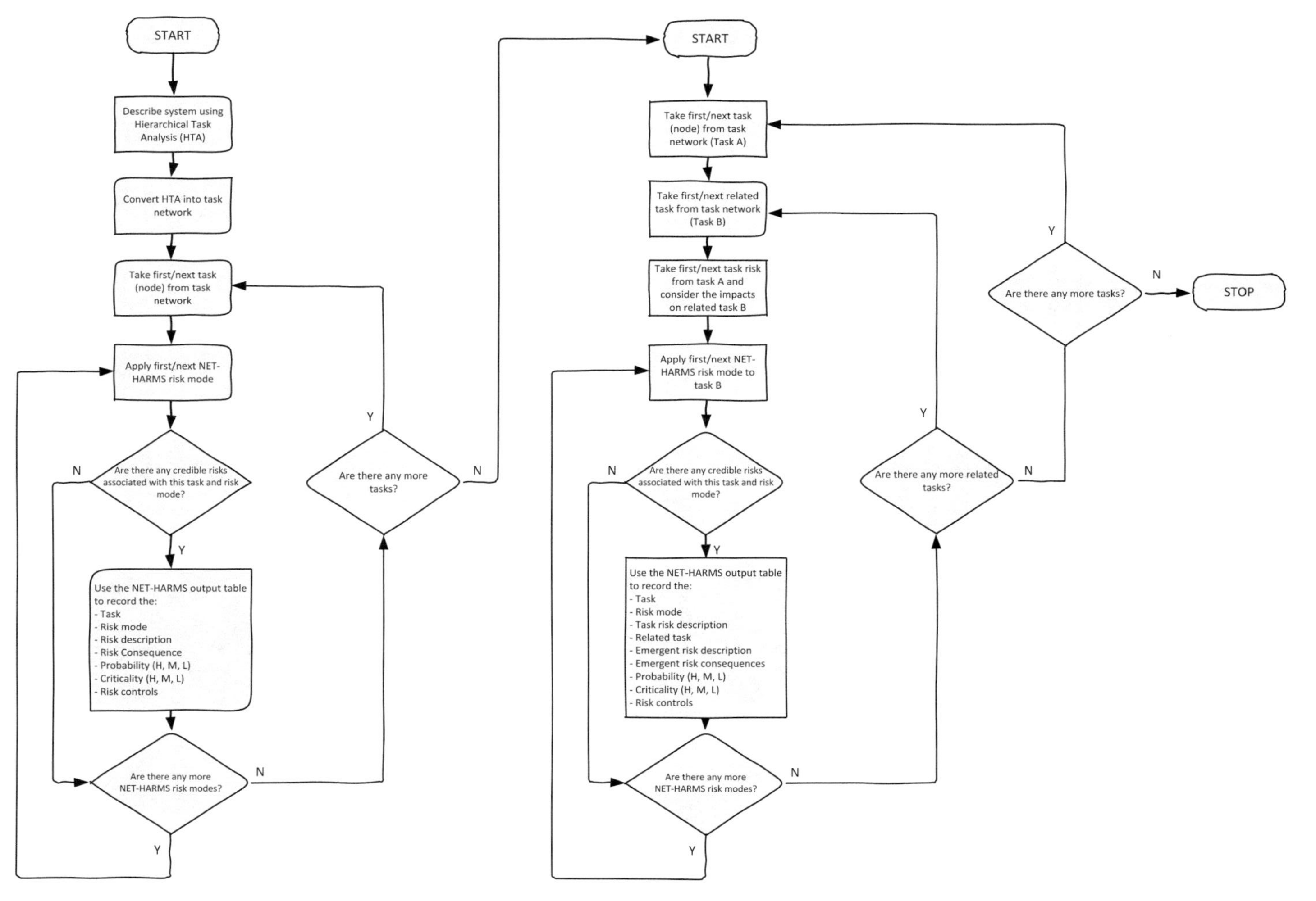

FIGURE 10.1 Net-HARMS procedure.

Applying Net-HARMS involves first developing a HTA that describes the goals, sub-goals, and operations undertaken within the system under analysis (Annett et al., 1971; Stanton, 2006). The HTA is then converted into a task network which shows these activities in the form of a network, depicting the relationship between tasks.

Once the task network is finalised, task risks are identified by applying the Net-HARMS risk mode taxonomy (see Table 10.1) to each node within the task network. For risk modes that are deemed credible (i.e., could conceivably occur), the analyst provides a description of the risks and their consequences, ratings of their probability and criticality (low, medium, or high), and suggested risk controls (Salmon et al., 2022). Following task risk identification, the risk mode taxonomy is applied once more to identify emergent risks that are likely to arise during instances when the identified task risks occur and influence the conduct of other tasks (Salmon et al., 2022). This is a critical feature that attempts to support the identification and management of new and unexpected risks that are created when performance elsewhere is sub-optimal – a central feature of accident causation. This component of the analysis is based on the relationships between tasks described in the task network and involves assessing the emergent properties that could arise when risks from a related task influence the performance of the present task. For each credible emergent risk the analyst provides a description of the identified emergent risks and consequences, ratings of probability and criticality (low, medium, or high), and suggested risk controls. The outputs thus include an in-depth description of potential

TABLE 10.1 Net-HARMS risk mode taxonomy (adapted from Dallat et al., 2018)

Behaviour	*Risk modes*	*Description*
Task	T1 – Task mistimed	Task is undertaken too early or late within the process
	T2 – Task omitted	Task is not undertaken
	T3 – Task completed inadequately	Task is undertaken but is completed inadequately due to personal, environmental, or sociotechnical system influences
	T4 – Inadequate task object	The objects used to complete the task are inadequate in some way (e.g., poorly designed, malfunctioning, in poor condition, or inappropriate for the task
	T5 – Inappropriate task	An inappropriate task is performed instead of the required task
Communication	C1 – Information not communicated	The information required to complete the task is not communicated between system actors
	C2 – Wrong information communicated	The wrong information is communicated between system actors
	C3 – Inadequate formation communicated	Information is communicated between actors but is inadequate (e.g., incomplete or missing information, information is out of date)
	C4 – Communication mistimed	Communication of the required information is undertaken too early or late
Environmental	E1 – Adverse environmental condition	The task is undertaken in adverse environmental conditions that influence task performance

tasks and emergent risks which are then used to support the development of appropriate risk controls.

Applications in sport

Net-HARMS was originally developed for the outdoor activity sector (Dallat et al., 2018; 2023); however, the approach is generic in nature and can be applied in any domain for prospective risk assessment purposes. Since its initial applications in outdoor activity, Net-HARMS has only had one application in sport: to identify risks to performance in elite women's cycling (Hulme et al., 2021). However, Net-HARMS has potential to be applied to proactively identify risk in injury; performance of teams, clubs, leagues; and player transfers, among many others. Net-HARMS has been applied outside of sport for identifying the safety risks throughout a railway level crossing lifecycle (Salmon et al., 2022), the risks associated with hazardous manual handling tasks (McCormack et al., 2023), and risks to patient safety in medication administration (Salmon et al., 2023).

Procedure and advice

A flowchart depicting the Net-HARMS procedure is presented in Figure 10.1. Step-by-step guidance is presented below.

Step 1: Define aims of the analysis and the task or system under analysis

The first step in applying Net-HARMS involves clearly defining the aims and boundaries of the analysis. As described above, one of the key strengths of the Net-HARMS method is that it can be used to identify risks across overall systems (e.g., clubs or leagues), as opposed to risks only at the sharp end of system operation (e.g., player or coach activities). It is therefore important during this step to clearly define the system levels that will be considered during the analysis. When using Net-HARMS, it is recommended that the analysis should consider as much of the sociotechnical system as is possible within the project constraints. This can include levels beyond the sharp end, including the organisational level up to and including sports governing bodies, government, and international organisations. For example, based on the aims, resources, access to data, and time available, it may be only necessary to consider risks within a team, club, or organisation. Alternatively, on larger scale projects with more resources it may be appropriate to consider additional risks at the sports governance, government, and international levels of the sport system in question. In the community football case study example presented in this chapter, the boundary was set at the level of the football club, as the organisation was interested in risks that they could conceivably control.

Step 2: Construct a hierarchical task analysis for the system under analysis

Once the aims of the analysis and system boundary are clearly defined, a HTA should be created for the system under analysis. HTA (see Chapter 3) works by decomposing systems and behaviour into a hierarchy of goals, sub-ordinate goals, operations, and plans to describe "what an operator is required to do, in terms of actions and/or cognitive processes to achieve a system goal" (Kirwan & Ainsworth 1992). The HTA output specifies the overall goal of a system, the sub-goals required to achieve this goal, the operations required to achieve each of the sub-goals specified, and the plans that dictate the sequence in which operations are undertaken.

Initially, data should be collected to support the development of a valid HTA. This should include data regarding system goals, sub-goals, and the operations that are undertaken to achieve them. Further, information regarding the ordering of tasks, and the factors that influence behaviour, task performance, and task outcome should be collected where possible. A number of different approaches can be used to collect this data, including observations, concurrent verbal protocols, SME workshops, structured or semi-structured interviews, e.g., the critical decision method (Klein et al., 1989), questionnaires and surveys, walkthrough analysis and documentation review (e.g., incident reports, standard operating procedures). The data collection approaches used are dependent upon project constraints, such as time, resources, access, and the number of analysts available. For detailed guidance on how to develop the HTA, the reader is referred to Chapter 3.

Step 3: Create task network

The next step involves converting the HTA into a task network which is then used as the basis on which to identify task and emergent risks. Task networks (see Figure 10.2), originally proposed as part of the EAST framework (see Chapter 5), are used to represent interrelated system tasks in the form of a network (Stanton et al., 2013). This enables analysts to understand the interactions and coupling that exists between tasks across the system. Within Figure 10.2 the circular nodes represent tasks and the arrows linking the tasks represent relationships between tasks. Relationships between tasks are included in the task network when:

1. Tasks are undertaken sequentially, e.g., task 3 is undertaken after completion of task 2;
2. Task are undertaken together, e.g., tasks 1 and 2 are undertaken together;
3. The outcomes of one task influence the conduct of another, e.g., the outcomes of task 4 influence how task 5 is undertaken; or
4. The conduct of one task is dependent on completion of the other, e.g., task 5 cannot be undertaken until task 4 is completed.

Task networks are constructed by taking the first layer of sub-goals from the HTA and identifying which of them are related to one another based on the relationship types described above. For example, in the community football club example presented later, the tasks 'coaching' and 'play football' are linked as coaching influences how the team play in matches. Similarly, the task 'recruit and manage sponsors' is linked to the task 'finances and accounting' as sponsorship influences club finances.

Once the initial draft task network is complete it is useful to have suitable SMEs review it, including both the nodes included in the network and the relationships between nodes. The task network should then be refined based on the SMEs feedback. It is normal practice for the task network to go through various iterations before it is finalised.

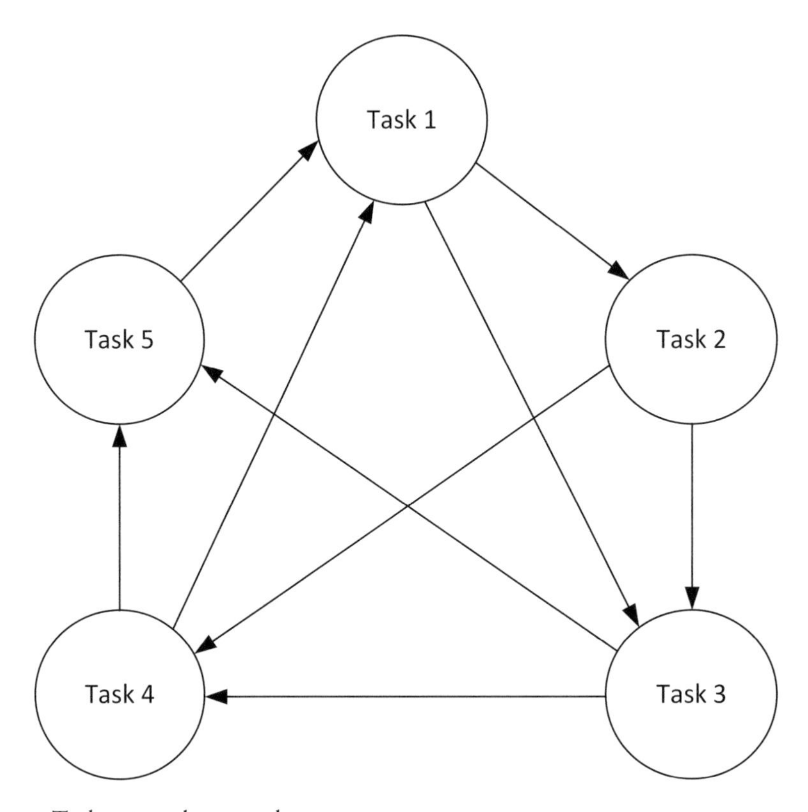

FIGURE 10.2 Task network example.

Step 4: Identify task risks

The task network is first used to support the identification of 'task risks' that could potentially emerge during the conduct of each task. Task risks are defined as those risks that could emerge if tasks are not completed optimally (Salmon et al., 2022). Here the analyst takes each task from the task network and uses the Net-HARMS risk mode taxonomy (see Table 10.1) to determine whether there are any credible risks. All ten risk modes should be applied to each task, with the analyst considering what risks could emerge if the risk modes are deemed to be credible. For example, for the community football case study example, the analyst may determine that the risk mode 'T3 – task completed inadequately' is credible for tasks such as 'strategic and operational planning', 'manage player health and wellbeing', and 'coach recruitment'. For each identified risk, the analyst should give a description of the form that the risk would take, such as:

- 'Strategic and operational planning is undertaken inadequately due to issues such as time constraints, lack of supporting information, or absence of key personnel'
- 'Management of player health and wellbeing is undertaken inadequately due to a lack of formal processes'
- 'Coach recruitment is undertaken inadequately due to time pressures and financial constraints'.

TABLE 10.2 Net-HARMS risk types

Risk category	*Description*
Reputation	Refer to risks that may adversely impact the reputation of stakeholders associated with the task or system under analysis
Financial	Refer to risks that may adversely impact the financial status of stakeholders associated with the task or system under analysis
Safety	Refer to risks that may adversely impact the safety of stakeholders associated with the task or system under analysis
Performance	Refer to risks that may adversely impact the performance the task or system under analysis
Legal	Refer to any identified legal risks for stakeholders associated task or system under analysis

For each credible risk, analysts should then determine and describe the associated consequences. For the risk 'strategic and operational planning is undertaken inadequately due to issues such as time constraints, lack of supporting information, or absence of key personnel' the consequence recorded would be 'club develop and implement a sub-standard strategy and operational plan'. For the risk of 'Management of player health and wellbeing is undertaken inadequately due to a lack of formal processes', the consequence would be 'player health and wellbeing is not adequately managed, increasing the risk of poor player health and wellbeing'. For the risk of 'coach recruitment is undertaken inadequately due to time pressures and financial constraints', the consequence would be 'club recruit a coach that does not align with values, strategy, vision, and philosophy'. The level of description provided for each risk is dependent on the time available and the analyst's knowledge of the domain; however, it is recommended that the analyst at least describes the risks that directly impact on performance.

Once the consequence is described, analysts should then provide a rating of the probability of the risk occurring. An ordinal probability scale of low, medium, or high is typically used. If the risk has not occurred previously, or the analyst deems it to be unlikely, a low (L) probability is assigned. If the risk has occurred on previous occasions, or the analyst determines that it could occur in certain situations, a medium (M) probability is assigned. Finally, if the risk has occurred on frequent occasions, or the analysts determines that it is likely to occur frequently, a high (H) probability is assigned.

Next, the analyst assigns a criticality rating for each of the task risks, again using scale of low, medium, and high. Here the analyst should consider how critical the risk is to the overall task/system under analysis. Normally, if the risk would lead to a critical incident or large-scale failure, it is rated as a highly critical risk.

A final optional phase of the task risk identification step involves classifying each of the identified task risks into one of the risk categories described in Table 10.2.

Step 5: Identify emergent risks

Once the analyst is satisfied that all task risks have been identified, the emergent risk identification phase can begin. Emergent risks represent new risks that arise as a result of the interaction between the task risks identified during Step 4 (see Figure 10.3) and other

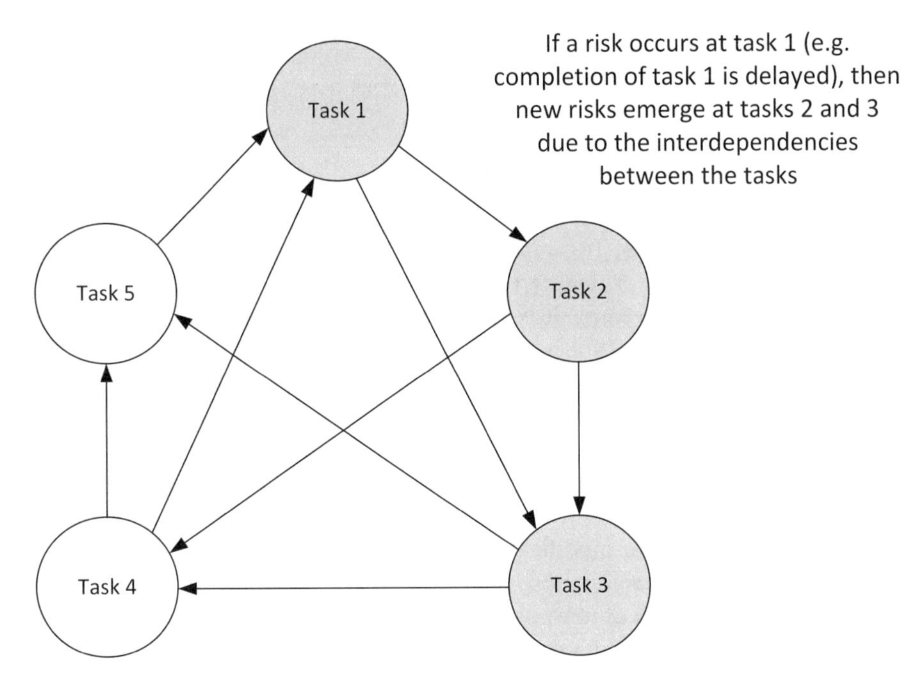

FIGURE 10.3 Emergent risk example.

tasks in the task network. Identifying emergent risks involves identifying which tasks are related via the task network, and then identifying what new risks emerge when task risks from one task influence a related task. For example, if tasks A and B are related, then analysts should determine what the impact on task B is if task A is undertaken inadequately due to the presence of a task risk from step 3 (Salmon et al., 2022). The Net-HARMS taxonomy is used again to support the identification of emergent risks. For example, if the analyst determines that a task risk for task A will result in task B being delayed, then they should classify the emergent risk as 'T1 – task mistimed'.

For example, the task of 'coach recruitment' is linked in the task network to 'coaching'. For the task risk 'coach recruitment is undertaken inadequately due to time pressures and financial constraints', the analyst could identify an emergent risk at the 'coaching' task of 'coaching practices are inadequate', as it is likely that the coach recruited via an inadequate recruitment process would not deliver a coaching philosophy that is consistent with club values, strategy, vision, and philosophy.

For each credible emergent risk, analysts should determine and describe the associated consequences. As with the task risk phase, analysts then provide a rating of the probability of the emergent risk occurring. An ordinal probability scale of low, medium, or high is again used. It is worth noting that the probability of emergent risks is often rated as high as the initial task risk is viewed as having occurred, which in turn increases the likelihood that the emergent risk will occur also.

Next, the analyst assigns a criticality rating for each of the risks. Again, a scale of low, medium and high is used to rate risk criticality. Here the analyst should consider how critical the risk is to the overall task/system under analysis. Normally, if the risk would lead to a critical incident (in relation to the task/system in question), it is rated as a highly critical risk.

The final phase of the emergent risk identification step involves classifying each of the identified emergent risks into one of the risk categories described in Table 10.1.

Step 6: Identify risk controls

The final stage of the Net-HARMS process involves proposing risk controls for the identified task and emergent risks. If there are time constraints, it is useful to focus only on the high probability high criticality risks on the basis that these represent at a minimum the risks that should be controlled by the stakeholders involved. It is recommended that a group of analysts or SMEs be used to identify appropriate risk controls. The group should work through the task and emergent risk tables and discuss risk controls that can be used to either prevent the risk from occurring or mitigate the consequences once the risk occurs. According to Stanton (2005) risk controls proposed following risk assessment applications are normally grouped under the following four categories:

1 equipment (e.g., redesign or modification of existing equipment);
2 training (e.g., changes in training provided);
3 procedures (e.g., provision of new, or redesign of old, procedures); and
4 organisational (e.g., changes in organisational policy or culture).

Step 7: Review and refine analysis

Once the Net-HARMS analysis is complete it is useful to have various SMEs review it to determine the credibility of the risks identified, review the probability and criticality ratings, and confirm that the proposed controls are appropriate. The analysis should then be refined based on SME feedback. This typically involves removing identified risks, identifying new risks, modifying probability and criticality ratings, and adding or refining risk controls.

Advantages

- Net-HARMS goes beyond existing sharp-end focused risk assessment methods to identify risks across overall systems;
- Previous applications demonstrate the capacity of Net-HARMS to identify the risks to elite sports performance (e.g., Hulme et al., 2021);
- Net-HARMS identifies both task risks and emergent risks. The latter is a significant advancement over existing risk assessment methods (Dallat et al., 2019);
- The analysis is comprehensive, identifying potential risks across the overall system, including sports club management, governance, and international levels;
- Net-HARMS analyses provide recommended risk controls for identified task and emergent risks;
- The Net-HARMS taxonomy prompts the analyst for potential task and emergent risks (this goes beyond typical brainstorming and experienced-based risk assessment processes);
- Net-HARMS is based on SHERPA, which, according to the literature has the most positive reliability and validity evidence of existing human error identification methods (Stanton et al., 2013);

- Net-HARMS is easy to learn and apply and requires minimal training;
- The Net-HARMS method and taxonomy are generic, allowing it to be used in any domain;
- Conducting the HTA and task network analyses enables analysts to develop an in-depth understanding of the system under analysis;
- The outputs effectively create a risk register detailing the various possible risks within the system under analysis;
- Net-HARMS goes beyond safety risks to also identify reputational, financial, performance, and legal risks.

Disadvantages

- Net-HARMS can be time-consuming to apply, particularly when the analysis boundary includes organisational, regulatory, and government stakeholders;
- Net-HARMS is a relatively new method with few published applications compared to more established methods;
- The quality of the analysis is highly dependent on the expertise of the analysts involved;
- The analysis incurs high levels of repetition, with risks often being identified multiple times (while this is useful for comprehensiveness it can be laborious for the analysts involved);
- Initial reliability and validity studies have found only moderate levels of reliability and validity (e.g., Hulme et al., 2021; Salmon et al., 2023);
- There is currently no performance shaping factors taxonomy to support consideration of contextual factors that may increase the likelihood of task and emergent risks.

Related methods

Net-HARMS requires an initial HTA of the system under analysis. The HTA is subsequently used to develop a task network for the system under analysis, a component that is derived from the EAST framework (see Chapter 5). The Net-HARMS approach and taxonomy was developed based on the SHERPA human error identification method (Embrey, 1986).

Approximate training and application times

Net-HARMS is a relatively simple method that requires little training. In a recent reliability and validity study participants were provided with training that lasted around two hours and were able to achieve moderate levels of validity (Hulme et al., 2021). Application time depends on the system under analysis and the size and interrelatedness of the task network; however, Net-HARMS analyses typically require significant time as the analysis goes beyond the sharp end to consider organisational, regulatory, and government activities. For large and complex task networks the application can extend to weeks; however, for smaller and simpler analyses it is possible to complete Net-HARMS analyses in one to two days.

Reliability and validity

The SHERPA method has a significant body of promising validation evidence associated with it (Stanton & Stevenage 1998; Stanton & Young 1999; Stanton et al., 2013). Hulme et al. (2021) compared the reliability and validity of Net-HARMS, STPA (see Chapter 9) and EAST-BL (see Chapter 5). Participants received training in either Net-HARMS, STPA, and EAST-BL and were asked to use their respective method to identify the risks that could emerge throughout a railway level crossing design lifecycle. Each participant's identified risks were then compared with a gold standard expert analysis and the signal detection theory (SDT) paradigm (Green & Swets, 1966) was used to calculate validity statistics. Hulme et al. (2021) found a weak to moderate positive correlation coefficient for Net-HARMS, meaning that the method did indeed support relatively novice analysts with the identification of credible system risks. Salmon et al. (2023) also tested the criterion-referenced validity of Net-HARMS when used by healthcare practitioners to identify potential risks associated with a patient medication administration task. According to Salmon et al. (2022) Net-HARMS achieved poor levels of validity for both task and emergent risk identification. This was attributed to a shorter training period than Hulme et al. (2021) as well as analyst inexperience.

Tools needed

Net-HARMS can be conducted using pen and paper; however, analyses are normally undertaken using a Microsoft Excel spreadsheet. The HTA tool can be used for the HTA development component and task networks are typically drawn in Microsoft Visio or PowerPoint.

Case study example: community club football

A local grassroots football club was seeking to modify its operations to ensure its long-term sustainability as a community football club. As part of their efforts to strengthen financial sustainability, the club proposed to run two new football programmes, and acquire management of critical club facilities from the local council. These proposals were focused on broadening the club's engagement with community, and diversifying participation in football via walking and disability football programmes. To secure more control over club operations, the club was seeking to take over the lease of the playing facilities, currently held by the local Council.

It was envisaged that the proposed changes would create a shift from the club's traditional way of operating; therefore, at the time of the study it was critical to understand how the new programmes and facility management activities would impact operations, and what risks may arise. The present Net-HARMS analysis was undertaken to enable decisions to be made regarding an effective and appropriate implementation of the proposed club restructure with appropriate controls in place to manage operational, financial, and reputational risks.

Methods

The community football club Net-HARMS analysis involved the development of a task network for the club's operations followed by an application of the Net-HARMS

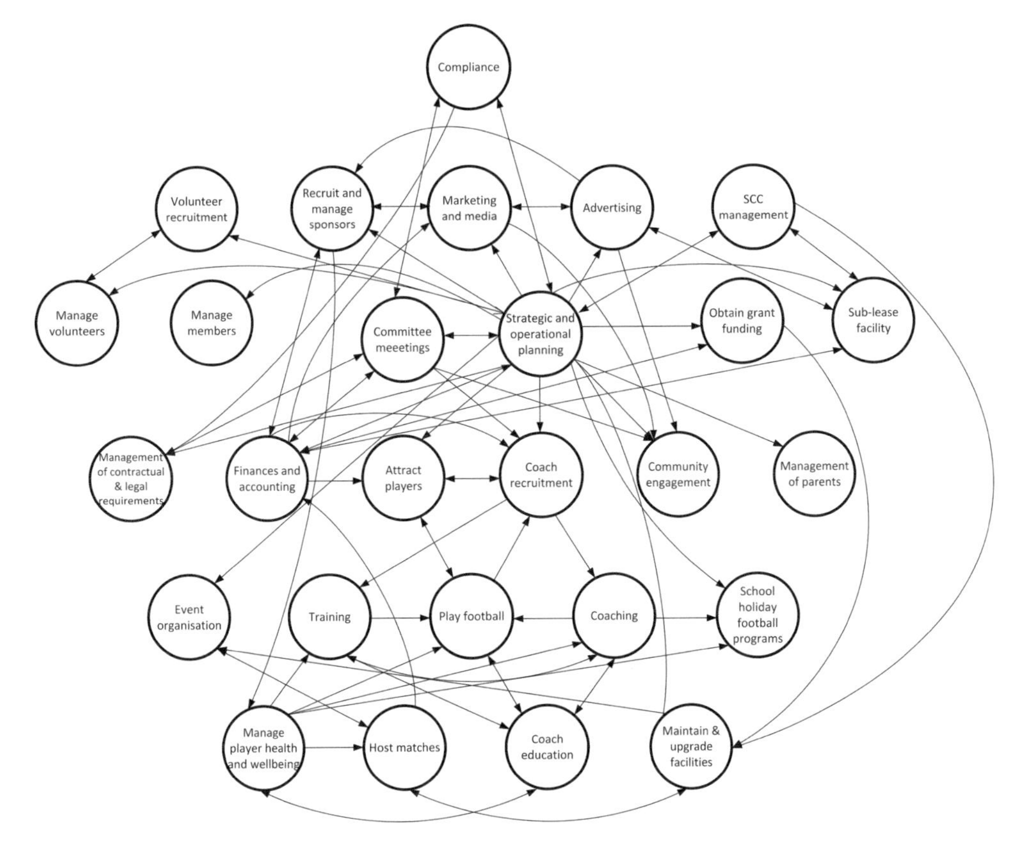

FIGURE 10.4 Football club operations task network.

taxonomy to identify task risks. The authors first developed a task network for the club's activities including the proposed new programmes and lease takeover tasks (see Figure 10.4). The task network and Net-HARMS taxonomy were then used to identify potential task risks. Specifically, one analyst used the Net-HARMS taxonomy to identify credible risks that could occur during each of the 27 identified tasks. For each credible task risk, the analyst provided a description of the risk, its consequences, and ratings of probability and criticality. The identified risks were subsequently categorised as either reputational, financial, safety, performance, or legal based on the impact that they would have on club functioning.

Risks identified

A total of 179 task risks were identified. The number of risks associated with each task from the task network is presented in Figure 10.5.

The tasks with the most risks associated with them included facility upgrades, facility maintenance, and school holiday football camps, all with ten associated risks. Next are manage player health and wellbeing, committee meetings, event organisation and financial management and accounting, each having nine associated risks. All tasks within the task network had credible risks associated with them.

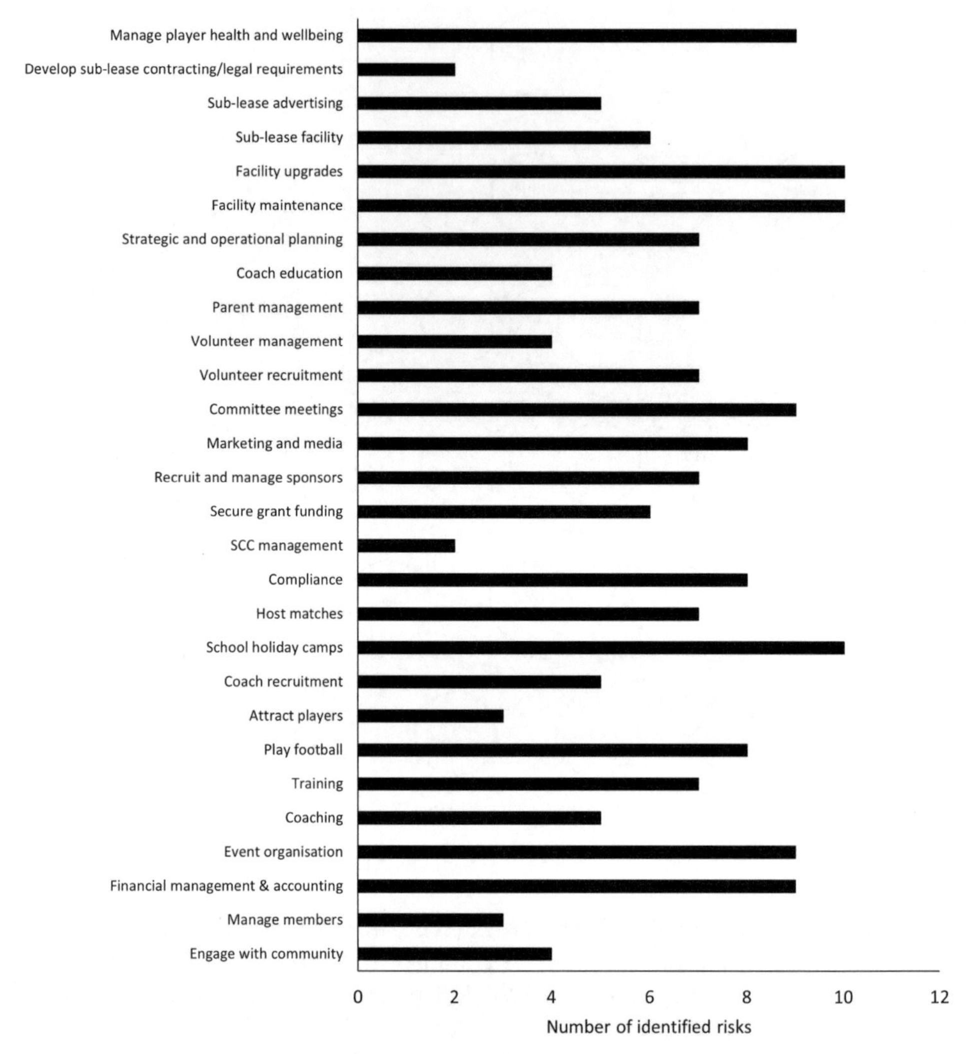

FIGURE 10.5 Number of risks associated with each task from the football club operations task network.

Risk types

The breakdown of risk types is presented in Figure 10.6.

As shown in Figure 10.6 reputational risks were the most frequently identified risk type (n = 146). Financial risks were the next most frequent (n = 138), followed by performance risks (n = 79), safety risks (n = 54) and legal risks (n = 54).

High probability high criticality risks

Of the 179 risks identified, 22 were classified as both high probability and high criticality (Table 10.3). The findings show that the high probability high, high criticality risks relate to community engagement, event organisation, facility management, school holiday football camps, compliance activities, sub-leasing of facilities, contracts, and the

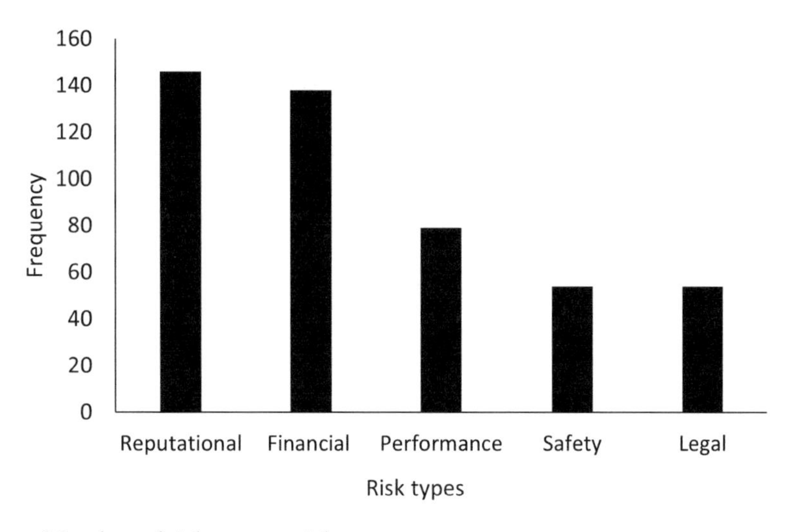

FIGURE 10.6 Number of risks across risk types.

TABLE 10.3 High probability high criticality risks

Tasks	*Task risk description*	*Task risk consequences*	*P*	*C*	*Form of risk*
Engage with community	Community engagement activities are inadequate	Local communities are not engaged	H	H	Reputational, financial
Event organisation	Event organisation activities are delayed	Club events are not well organised	H	H	Reputational, financial
Event organisation	Event organisation activities are inadequate	Club events are not well organised	H	H	Reputational, financial
Play football	Pitches are in poor condition	Game plan cannot be implemented, and injury risk is enhanced	H	H	Performance, reputational, safety
School holiday camps	Planning and preparation for school holiday camps is delayed	School holiday camps are not successful (e.g., low numbers, inadequate programme, poor experience)	H	H	Reputational, financial
School holiday camps	School holiday programme advertising is delayed	School holiday programme is not successful due to low numbers	H	H	Reputational, financial
Host matches	Match objects are inadequate, e.g., pitches are in poor condition	Players/Coaches/Members have poor experience during match events	H	H	Reputational, financial
Compliance activities	Compliance activities are not undertaken as required	Club is not compliant with relevant football governance	H	H	Legal, reputational, financial
Compliance activities	Compliance activities are inadequate	Club is not compliant with relevant football governance	H	H	Legal, reputational, financial

(*Continued*)

TABLE 10.3 (Continued)

Tasks	*Task risk description*	*Task risk consequences*	*P*	C	*Form of risk*
Sub-lease facility	Sub-leasing is not achieved due to issues, such as poor advertising and failure to maintain facilities	Earning potential associated with sub-leasing is not fulfilled	H	H	Financial, reputational
Sub-lease facility	Sub-leasing is inadequate due to issues, such as poor advertising and failure to maintain facilities	Earning potential associated with sub-leasing is not fulfilled	H	H	Financial, reputational
Sub-lease facility	Facilities are not maintained and so are not fit for sub-leasing	Earning potential associated with sub-leasing is not fulfilled	H	H	Financial, reputational
Sub-lease facility	Potential opportunities for sub-leasing are not communicated widely	Earning potential associated with sub-leasing is not fulfilled	H	H	Financial, reputational
Sub-lease facility	Communication of sub-leasing opportunities is delayed	Key opportunities for sub-leasing are missed	H	H	Financial, reputational
Sub-lease advertising	Advertising re sub-leasing opportunities is delayed	Key opportunities for sub-leasing are missed	H	H	Financial, reputational
Sub-lease advertising	Sub-leasing opportunities are not advertised	Earning potential associated with sub-leasing is not fulfilled	H	H	Financial, reputational
Sub-lease advertising	Advertising materials for sub-leasing opportunities are sub-standard	Earning potential associated with sub-leasing is not fulfilled	H	H	Financial, reputational
Develop sub-lease contracting/ legal requirements	Contracts are delayed	Sub-leasing is delayed with associated financial and reputational impacts	H	H	Legal, financial
Develop sub-lease contracting/ legal requirements	Contracts are sub-standard	Sub-leasing arrangements carry risk to the football club	H	H	Legal, financial

(*Continued*)

TABLE 10.3 (Continued)

Tasks	*Task risk description*	*Task risk consequences*	*P*	*C*	*Form of risk*
Manage player health and wellbeing	Injury and illnesses are managed inadequately (e.g., due to lack of appropriate training)	Player health and wellbeing is not managed resulting in serious injury/illness	H	H	Safety, legal
Manage player health and wellbeing	First aid equipment is inadequate	Player health and wellbeing is not managed resulting in serious injury/illness	H	H	Safety, legal
Manage player health and well being	Information regarding first aid procedures/ practices is not communicated to club volunteers	Player health and wellbeing is not managed appropriately resulting in serious injury/ illness	H	H	Safety, legal

Note: **P** – denotes probability, **C** – denotes criticality, **H** – indicates high.

management of player health and wellbeing (including during the school holiday football camps, walking, and disability football programmes). A majority of the high probability high criticality risks are reputational and financial risks, with the remaining risks being legal, financial, and safety-related.

Discussion

The Net-HARMS risk assessment revealed a series of risks that could arise during future football club operations. In particular, the risks that were classified as high probability and high criticality represent a set of potential risks that should be considered. These included risks associated with the tasks of community engagement, event organisation, facility management, school holiday camps, compliance activities, sub-leasing of facilities, contracts, and management of player health and wellbeing during all programmes. Based on the analysis presented, recommendations presented to the football club were that they review the risks and consider existing risk controls and whether any new controls are required.

The majority of the risks identified were reputational and financial in nature, with comparatively less of the identified risks relating to football performance, player, coach and volunteer safety, and legal aspects of club operations. While criticality of managing safety risks goes without saying, based on the analysis it was recommended that the club also consider specifically the reputational and financial risks when planning and undertaking club processes and functions. This finding demonstrates the capacity for Net-HARMS to identify risks beyond safety risks.

An aim of the Net-HARMS analysis was to identify risks associated with the proposed lease take over. The findings showed that there are a series of risks associated with maintaining and upgrading the facilities and sub-leasing the facilities. It was therefore recommended that the club develop, and document appropriate plans regarding the maintenance of facilities prior to the lease take over. Based on the risks identified, this plan should include maintenance and upgrade timelines, the acquisition of appropriate maintenance equipment, the training of volunteers in maintenance and upgrade tasks, and communication and feedback mechanisms regarding the status and maintenance needs of facilities.

In regard to sub-leasing the facility, the risks identified largely related to a failure to capitalise on the opportunity to sub-lease through failures in advertising and communication, facility maintenance and contract management. It was thus recommended that a sub-leasing facility plan be developed which incorporates specific plans around advertising and communication, facility maintenance, and contract management.

The risks identified via the Net-HARMS analysis provided a risk register detailing a set of risks that should be managed during community football club operations. The risk register should be used to support the development of risk controls and risk management practices. Further, the risk register will be required to be reviewed and updated periodically as part of risk management practices (i.e., as new risks emerge or as risks are no longer credible).

Recommended Reading

Dallat, C., Salmon, P. M., & Goode, N. (2018). Identifying risks and emergent risks across socio-technical systems: The NETworked hazard analysis and risk management system (NET-HARMS). *Theoretical Issues in Ergonomics Science*, 19(4), 456–482.

Dallat, C., Salmon, P. M., & Goode, N. (2019). Risky systems versus risky people: To what extent do risk assessment methods consider the systems approach to accident causation? A review of the literature. *Safety Science*, 119, 266–279.

Stanton, N. A. (2006). Hierarchical task analysis: Developments, applications, and extensions. *Applied Ergonomics*, 37, 55–79.

References

Annett, J., Duncan, K. D., Stammers, R. B., & Gray, M. J. (1971). *Task analysis. Department of Employment Training Information Paper* 6. London: Her Majesty's Stationary Office (HMSO).

Dallat, C., Salmon, P. M., & Goode, N. (2018). Identifying risks and emergent risks across sociotechnical systems: The NETworked hazard analysis and risk management system (NET-HARMS). *Theoretical Issues in Ergonomics Science*, 19(4), 456–482.

Dallat, C., Salmon, P. M., & Goode, N. (2019). Risky systems versus Risky people: To what extent do risk assessment methods consider the systems approach to accident causation? A review of the literature. *Safety Science*, 119, 266–279.

Dallat, C., Salmon, P. M., & Goode, N. (2023). Testing the validity of the networked hazard analysis and risk management system (Net-HARMS). *Human Factors and Ergonomics in the Manufacturing & Service Industries*, 33(4), 299–311.

Embrey, D. E. (1986). SHERPA: A systematic human error reduction and prediction approach. In *Proceedings of the international topical meeting on advances in human factors in nuclear power systems*. Knoxville, Tennessee.

Green, D. M., & Swets, J. A. (1966). *Signal detection theory and psychophysics*. New York: Wiley New York.

Hulme, A., McLean, S., Dallat, C., Walker, G. H., Waterson, P., Stanton, N. A., & Salmon, P. M. (2021). Systems thinking-based risk assessment methods applied to sports performance:

A comparison of STPA, EAST-BL, and Net-HARMS in the context of elite women's road cycling. *Applied Ergonomics*, 91, 103297.

Hulme, A., Stanton, N. A., Walker, G. H., Waterson, P., & Salmon, P. M. (2022). Testing the reliability and validity of risk assessment methods in human factors and ergonomics. *Ergonomics*, 65(3), 407–428.

Kirwan, B., & Ainsworth, L. K. (1992). *A guide to task analysis*. London: CRC Press.

Klein, G. A., Calderwood, R., & Macgregor, D. (1989). Critical decision method for eliciting knowledge. *IEEE Transactions on Systems, Man, and Cybernetics*, 19(3), 462–472.

McCormack, P., Read, G. J., Hulme, A., Lane, B. R., McLean, S., & Salmon, P. M. (2023). Using systems thinking-based risk assessment methods to assess hazardous manual tasks: a comparison of Net-HARMS, EAST-BL, FRAM and STPA. *Ergonomics*, 66(5), 609–626.

Salmon, P. M., Stanton, N. A., Walker, G. H., Hulme, A., Goode, N., Thompson, J., & Read, G. J. (2022). *Handbook of systems thinking methods*. Boca Raton, FL: CRC Press.

Salmon, P. M., & McLean, S. (2020). Complexity in the beautiful game: Implications for football research and practice. *Science and Medicine in Football*, 4(2), 162–167.

Stanton, N. A. (2005). *Handbook of human factors and ergonomics methods: A practical guide for engineering and design*. Aldershot: Ashgate.

Stanton, N. A. (2006). Hierarchical task analysis: Developments, applications, and extensions. *Applied Ergonomics*, 37, 55–79.

Stanton, N. A., & Stevenage, S. V. (1998). Learning to predict human error: Issues of reliability, validity and acceptability. *Ergonomics*, 41(11), 1737–1756.

Stanton, N. A., & Young, M. S. (1999). What price ergonomics? *Nature*, 399(6733), 197–198.

Stanton, N. A., Salmon, N. A., & Walker, G. H. (2018). *Systems thinking in practice: The event analysis of systemic teamwork*. Boca Raton, FL: CRC Press.

Stanton, N. A., Salmon, P. M., Rafferty, L., Walker, G. H., Jenkins, D. P., & Baber, C. (2013). *Human factors methods: A practical guide for engineering and design*, 2nd Edition. Aldershot: Ashgate.

Svedung, I., & Rasmussen, J. (2002). Graphic representation of accident scenarios: Mapping system structure and the causation of accidents. *Safety Science*, 40(5), 397–417.

11

THE ACCIDENT MAPPING (ACCIMAP) METHOD

Background

In the world of professional sport where the line between success and failure is incredibly thin, there are a panoply of often overlooked factors that influence both success and failure.

The accident mapping (AcciMap) method provides a means to describe adverse events through the lens of Jens Rasmussen's Risk Management fFamework (RMF) (Rasmussen, 1997) (see Figure 11.1 for an adapted RMF to a sports context). AcciMap was developed to support the analysis of adverse events from a systems thinking perspective whereby 'upstream' contributory factors from organisational, regulatory, and government processes could be mapped onto a system hierarchy to show how they create 'downstream' issues relating to front-line workers. (Svedung & Rasmussen, 1997). In line with Rasmussen's RMF, AcciMap is based on the principle that system functioning, behaviour, safety, and adverse events are emergent properties of complex sociotechnical systems created by the decisions and actions of all stakeholders within a system. In a sport context these stakeholders include governing bodies, leagues, clubs, coaches, and support personnel, not athletes alone (Figure 11.1) (Dodd et al., 2024). Optimal and sub-optimal system functioning is therefore the shared responsibility of all sport system stakeholders. As well as analysing failure events, AcciMap also presents a suitable approach for describing successful performance. For example, adverse events and successes in sport includes athlete injury, illness, abuse; performance failures of teams and clubs; successes of national teams and paradigm-shifting policy implementation, among many others. The AcciMap method enables such events to be analysed in-depth to understand how systemic factors interacted to enable them.

Rasmussen's framework makes a number of assertions regarding safety and system functioning. These have been modified below to fit the sports context:

1 Sports performance is impacted by the decisions and actions of all actors, not just front-line athletes, coaches, and support staff alone;
2 Both successes and failures in sport are caused by multiple contributing factors from across the sports system, not just a single poor decision or action from athletes or coaches;

DOI: 10.4324/9781003259473-14

RMF levels	Example stakeholders/actors
International influences	World Anti-Doping Agency International Olympic Committee
Governments and governing bodies	National governing bodies National governments
Regulatory bodies and associations	Media National Sporting Organisations Research groups
Teams and organisations	Sporting organisations Players union/association Sponsors
Direct supervisors, management, medical and performance personnel	Team doctor Sports scientist Dietician/Nutritionist Coach
Athlete, teammates, and opponents	Athletes Opponents Teammates Fans
Equipment and environment	Technology Data Weather Facilities

FIGURE 11.1 Rasmussen's RMF (adapted from Rasmussen, 1997; Naughton et al, 2024).

3. Adverse events can result from poor communication and feedback (or 'vertical integration') across levels of the sport system, not just from deficiencies at one level alone;
4. Lack of vertical integration is caused, in part, by lack of feedback within and across levels of the sport system;
5. Behaviours within sport systems are not static, they migrate over time and under the influence of various pressures such as performance, psychological, community, financial, geopolitical, and societal pressures;
6. Migration occurs at multiple levels of sport systems and is extremely difficult to monitor and track;
7. Migration of practices causes sport system defences to degrade and erode gradually over time, not all at once. Adverse incidents are caused by a combination of this migration and a triggering event(s).

Level				
International influences	International actors	International actors	International actors	
Governments and governing bodies	Governments & Governing body actors	Governments & Governing body actors	Governments & Governing body actors	
Regulatory bodies and associations	Regulatory bodies & association actors	Regulatory bodies & association actors	Regulatory bodies & association actors	
Teams and organisations	Team & organisation actors	Team & organisation actors	Team & organisation actors	
Direct supervisors, management, medical, and performance personnel	Supervisory actors	Supervisory actors	Supervisory actors	Supervisory actors
Athlete, teammates, opponents, and fans	Athlete, teammates, opponents, and fans	Athlete, teammates, opponents, and fans	Athlete, teammates, opponents, and fans	
Equipment and environemt	Equipment & environment actors	Equipment & environment actors		

FIGURE 11.2 Generic ActorMap.

Using AcciMap involves applying two analysis methods: the ActorMap and AcciMap. ActorMap forms the first analysis component and is used to provide a representation of the stakeholders ('actors') who undertake activities within the system and therefore share the responsibility for its performance. In a sport context, the ActorMap involves identifying relevant actors at the seven hierarchical levels in Figure 11.1 (though the levels used are flexible and can be modified if required). The resulting ActorMap shows the actors who make up the system and at which level they reside. While this is a useful system model in of itself, it also supports analysts in mapping incident contributory factors to the correct level of the system hierarchy when building the Accimap. To provide an example, a generic ActorMap is presented in Figure 11.2.

The AcciMap method forms the second analysis component and is used to identify and represent the network of contributory factors involved in the incident in question across the same system hierarchy as the ActorMap. Contributory factors are identified, mapped to one of the system hierarchy levels, and are then linked between and across levels based on cause-effect relations. A generic template for the AcciMap diagram is presented in Figure 11.3.

AcciMap is a flexible method that can be applied in any domain. To date, it has been applied to describe and analyse adverse events in multiple domains (e.g., Hulme et al., 2021; Salmon et al., 2020), as well as to analyse multiple incident datasets (McLean et al., 2020;

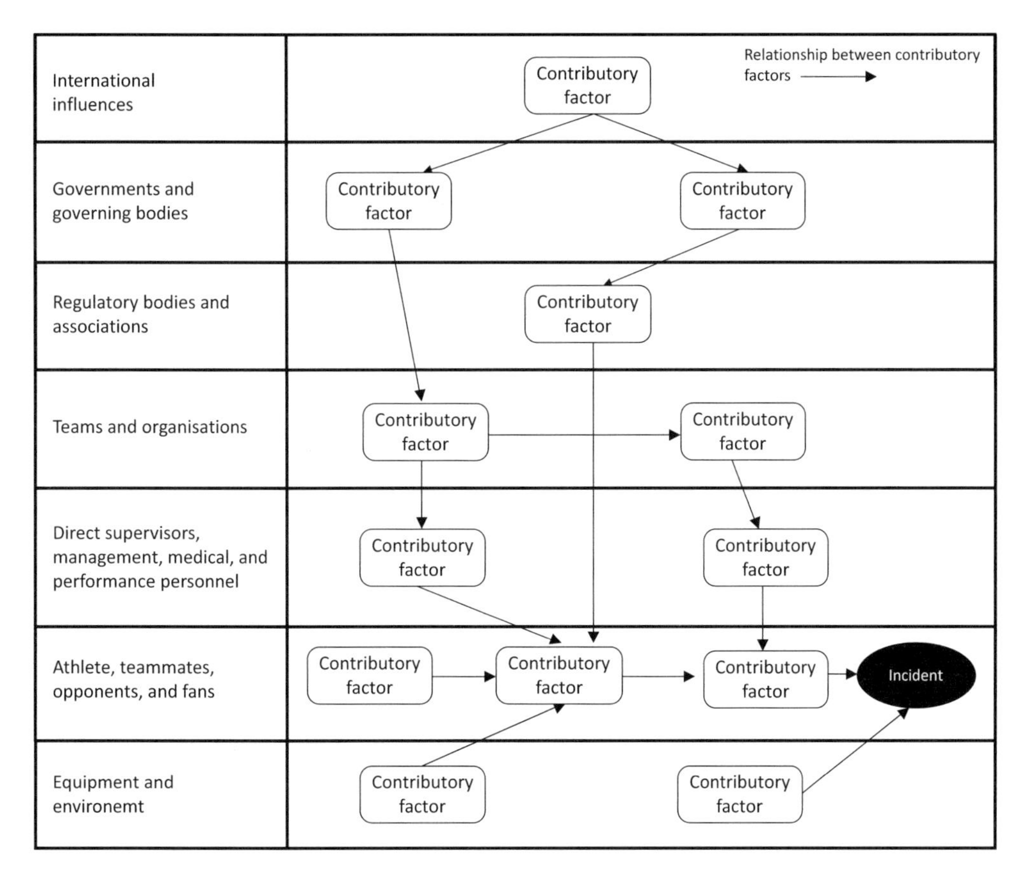

FIGURE 11.3 Generic AcciMap.

Salmon et al., 2014; 2017). In addition, outside of specific incidents AcciMap can also be used to examine recurring issues or a specific problem; for example, Salmon et al.'s (2022) analysis of work-related violence in hospitals, or Lane et al.'s (2020) AcciMap of the barriers preventing access to eating disorder treatment. This involves the use of SME workshops or systematic review to identify and map contributory factors across the hierarchical levels.

Applications in sport

AcciMap is a generic method and can be applied in any domain. The AcciMap method has not yet received widespread attention in sport; however, it has been used extensively in the area of outdoor recreation to examine injury, illness, and near-miss incidents during activities such as hiking, wheel sports, abseiling, swimming, and kayaking (e.g., McLean et al., 2021; 2022; Salmon et al., 2010; 2014; 2017). Applications in sport are emerging. For example, Dodd et al. (2024) used the RMF and AcciMap to visualise the enabling factors for child sexual abuse in sport. Further, Naughton et al. (2024) conducted a literature review on the contributory factors to doping in sport and subsequently mapped them to the RMF. There are numerous potential sports applications of AcciMap that could be used to provide a more detailed understanding of the factors that contribute to athlete injury, illness, and psychosocial incidents, as well as performance failures and successes. In the case study examples for this chapter we present AcciMaps of the factors that contributed to a single injury incident to an athlete, and aggregated set of injury, illness, and psychosocial incidents in outdoor recreation, and Eluid Kipchoge's assisted sub-2-hour marathon.

Procedure and advice

A flowchart depicting the AcciMap procedure is presented in Figure 11.4. Step-by-step guidance is presented below.

Step 1: Determine analysis aims and scope

The first step involves clearly defining the aims and scope of the analysis. Typically, the aim is to create a representation of the network of factors that interacted to create a particular adverse event (e.g., Salmon et al., 2012), or when using AcciMap to model successes, the factors that enabled success. Alternatively, the aim of the analysis may be to identify recurring contributory factors and relationships across multiple incidents derived from a dataset such as an incident reporting and learning system (e.g., McLean et al., 2021), or to identify the factors that interact to enable a recurring issue (e.g., Salmon et al., 2022), e.g., hamstring injuries in sport.

Defining the analysis boundaries will dictate how detailed the analysis should be in terms of the parts of the system hierarchy considered and how far back in time the analysis will go. It may be, for example, that an analysis boundary is set at the organisational level only, or may not extend to international factors outside of the country in which the incident occurred. Further, post-incident response may or may not be of interest (Salmon et al., 2022).

At a minimum, it is recommended that the analysis boundary incorporates levels above the organisation, including regulatory bodies, and government. Given the influence of international bodies in sport (e.g., WADA, IOC, FIFA) it is often useful to extend the boundary to consider international influences. For example, in their analysis of the factors that enable child sex abuse in sport, Dodd et al. (2024) included an international influences level to incorporate the influences of international sports bodies such as the International Olympic Committee and governing bodies such as FIFA.

Step 2: Data collection

AcciMap is dependent upon the collection of accurate data regarding the decisions, actions, and events that contributed to the incident(s) under analysis. This involves collecting data regarding the system and incident in question. Three broad forms of data are required (Salmon et al., 2022):

- Data on the sport system in terms of who shares the responsibility for safety and optimal system functioning.
- Data on the activities or processes that are undertaken within the sport system.
- Data on the incident(s) and any contributory factors that are perceived to have played a role in its occurrence.

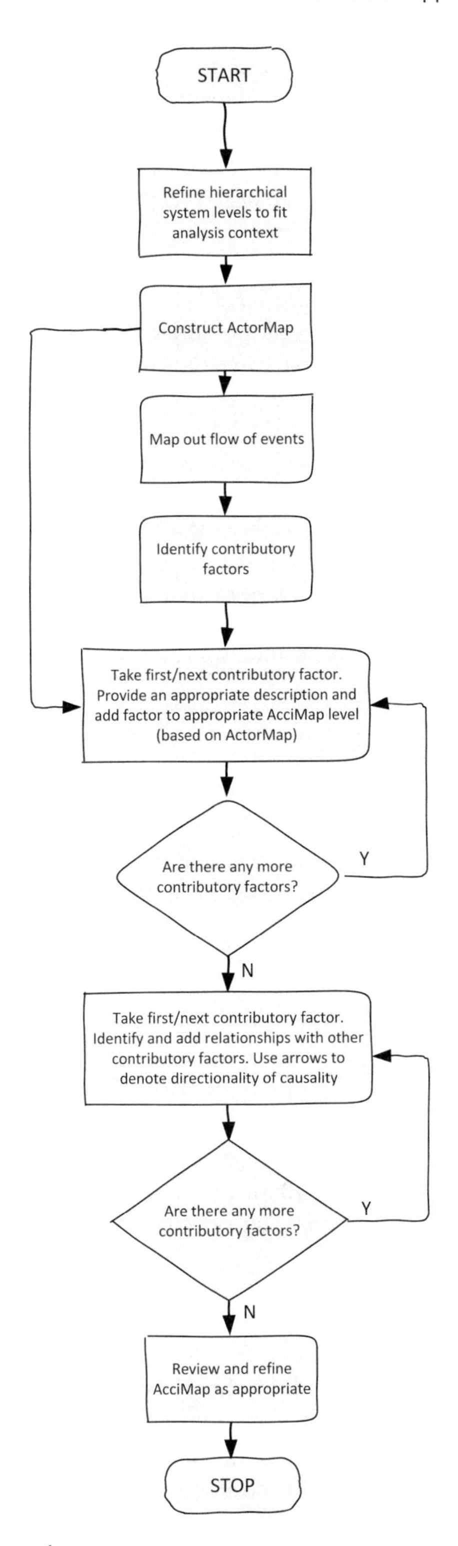

FIGURE 11.4 AcciMap procedure.

Salmon et al. (2022) outline a range of data collection activities to support the development of ActorMap and AcciMap analyses. These include interviews with those involved in the incident or SMEs for the domain, sport, or type of incident or event under analysis, reviewing reports or inquiries into the incident or event, and observing recordings of the incident or event. AcciMap analyses can be based on pre-existing data such as investigation reports (e.g., Newnam et al., 2017; Salmon et al., 2014), incident reporting and learning system data (e.g., Salmon et al., 2017), or data collected for the purpose of the analysis. The data collection process is typically iterative and often new data requirements are identified during the analysis process as potential contributory factors are unearthed. Ideally, multiple data collection approaches from those described above are used.

Step 3: Refine system hierarchy

Before constructing the ActorMap and AcciMap it is first useful to refine the hierarchical system levels to fit the system in which the analysis is taking place. This might include adding additional levels, such as an international level (e.g., Dodd et al., 2024; Naughton et al, 2024) or modifying the level descriptions to better fit the system under analysis (e.g., Salmon et al., 2022). In their analysis of the enabling factors for child sexual abuse in sport, Dodd et al. (2023) used the following levels:

- International influences
- Governments and governing bodies
- Regulatory bodies and associations
- Teams and organisations
- Direct supervisors, management, medical, and performance personnel
- Athletes, teammates, opponents, and fans Equipment and environment.

Step 4: Construct ActorMap

Once data collection is complete and the hierarchical system levels are agreed upon, the next step involves identifying all relevant actors who undertake activity within the system under analysis. Actors can be individual actors (e.g., athletes, coaches, support staff), artefacts (e.g., sports equipment, playing, and training facilities), or organisations (e.g., sports clubs, media, sponsors, governing bodies). Each actor should be annotated onto an ActorMap showing where across the sport system hierarchy they reside. An example ActorMap for the anti-doping system for rugby in Australia is presented in Figure 11.5.

Step 5: Map out the flow of events

When analysing individual events, it is useful to map out the flow of proximal events from left to right on the physical processes and activities level (second level from bottom) of the framework (Salmon et al., 2022). This initial timeline typically represents the proximal flow of events associated with actors who were directly involved in the incident. This flow of events should end at the right-hand side of the AcciMap with the adverse event outcome (e.g., athlete injury) (Figure 11.6).

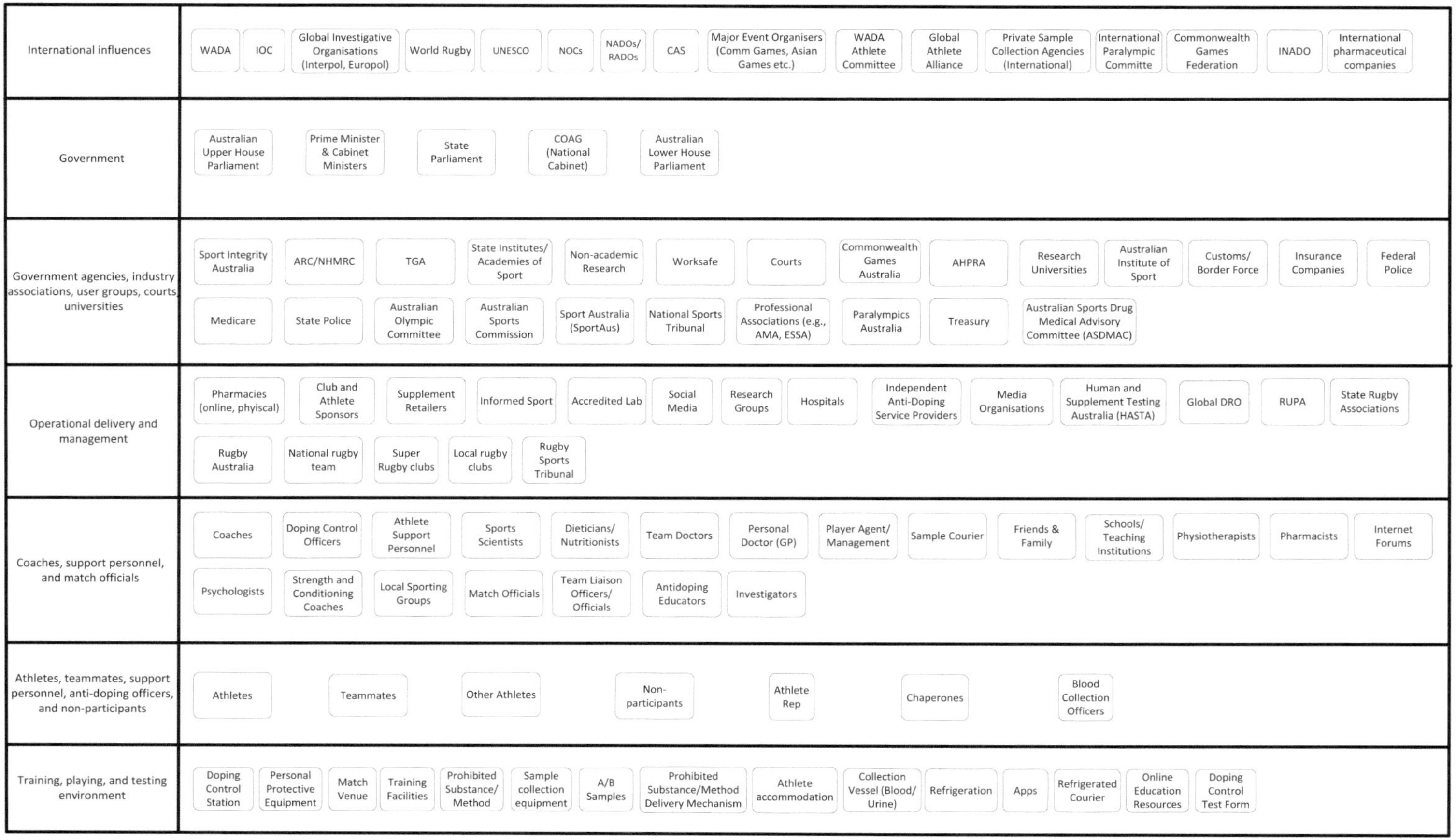

FIGURE 11.5 ActorMap of anti-doping stakeholders in Australian Rugby.

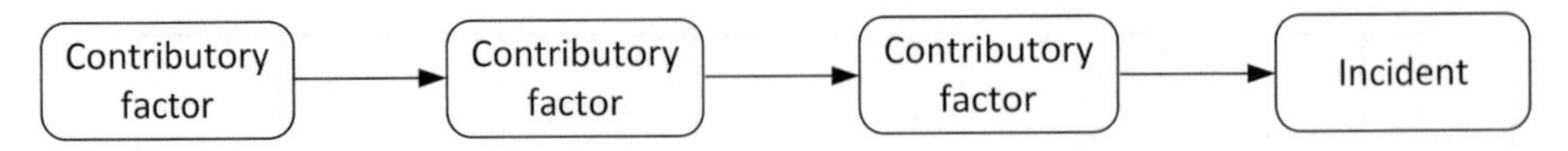

FIGURE 11.6 The flow of events from proximal events (left-hand side) to the incident (right-hand side) in the AcciMap.

Step 6: Identify contributory factors

Next the data should be analysed to identify contributory factors which played a causal role in the incident. This involves reviewing the data and recording any factors that the analyst feels played a contributory role in the events identified during Step 5. It is important to clearly define what is considered to be a contributory factor. According to Salmon et al. (2022), contributory factors can be defined as decisions, actions, and events:

> that, if it had not occurred or existed at the relevant time, then either the occurrence would probably not have occurred, adverse consequences associated with the occurrence would probably not have occurred or have been as serious, or another contributing safety factor would probably not have occurred or existed.
>
> *(Australian Transport Safety Bureau, 2008)*

When identifying contributory factors the analyst or team of analysts should take as broad a view as possible. As outlined in Salmon et al. (2022) a key requirement is to look for contributory factors across all levels of the system hierarchy and also contributory factors associated with all of the actors and organisations identified in the ActorMap.

Step 7: Map contributory factors on AcciMap

Once an initial set of contributory factors have been identified, the next step involves mapping them onto the AcciMap diagram. It is important here to use the ActorMap to inform this process since the ActorMap shows where different actors and organisations reside in the system and hence where contributory factors associated with each actor should be placed. For example, if an organisation at the teams and organisations level provided a sub-standard training programme, then the 'training programme' contributory factor should be placed at the teams and organisations level. To map the contributory factors, it is recommended that the analyst should take each contributory factor, identify which actor and organisation it is associated with, develop an appropriate description, and place the factor at the corresponding level on the AcciMap diagram. Contributory factor descriptions should use neutral language and try to avoid failure-based language where possible (Salmon et al., 2022). For example, for the training programme example above, a suitable description would be 'training programme' (rather than 'Inadequate training'). This process should continue until all contributory factors have been placed on the AcciMap. It is then useful to review the contributory factors and question whether there is a need to add additional factors that may link to those already identified.

Step 8: Identify and add relationships between contributory factors

Once all contributory factors have been identified the relationships between them should be identified and added to the AcciMap. This involves taking each contributory factor in turn and considering:

a whether it had an influence on any of the other contributory factors in the AcciMap; and
b whether it was influenced by any of the other contributory factors in the AcciMap.

According to Salmon et al. (2022), a relationship should be specified if one contributory factor creates, influences, or occurs in sequence with, another contributory factor. When a relationship is found the analyst draws a line to depict the relationship on the AcciMap, with an arrow showing the direction of influence. For example, the following contributory factors and relationships are presented later in the case study applying AcciMap for an injury incident:

- Repetitive stressors such as matches in a congested fixture period will influence athlete load (Figure 11.7).
- Match scheduling at a league level will influence travel bookings at a club level (Figure 11.8).
- At a club level, available funding will influence staffing which in turn will influence the competence of the staff member which in turn will influence decision quality (Figure 11.9).

The process of identifying the relationships between contributory factors also acts as a useful check on the contributory factors included in the AcciMap. If it is not possible to identify a relationship with any of the other contributory factors, the analyst should consider whether the contributory factor should be included within the AcciMap or whether further contributory factors need to be identified. As such, Steps 6–8 often occur in an iterative cycle until all relevant contributory factors and relationships have been identified.

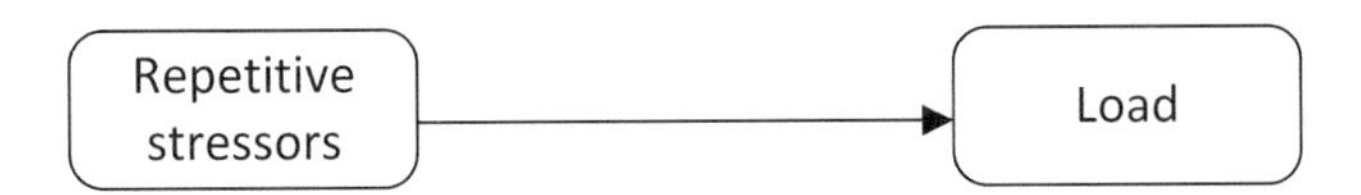

FIGURE 11.7 Relationship between repetitive stressors and load.

FIGURE 11.8 Relationship between match scheduling and travel bookings.

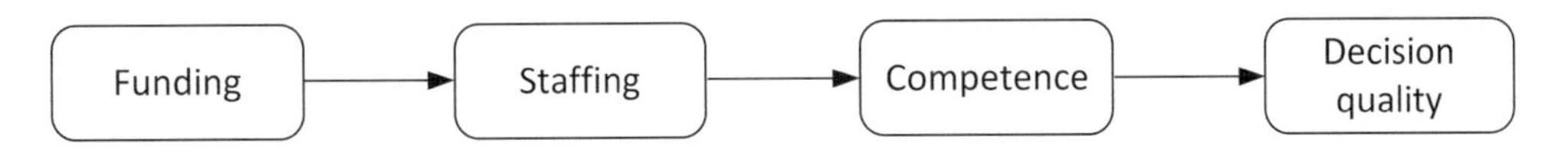

FIGURE 11.9 Relationship between funding, staffing, competence, and decision quality.

Step 9: Finalise and review AcciMap diagram

The output of Step 8 is a draft AcciMap diagram. The analyst(s) should next review the draft AcciMap and ensure that all contributory factors and relationships are included and that the descriptions are appropriate. It is useful during this step to return to the data and review it to verify the contributory factors and relationships identified. It is recommended that multiple reviews are undertaken by multiple analysts during this step.

Step 10: Subject matter expert review

The final stage of the process involves asking appropriate SMEs to review the final ActorMap and AcciMap diagrams. It is best practice to use SMEs who were either involved in the incident or event who have extensive knowledge of the system and activities under analysis (Salmon et al., 2022). The AcciMap should be updated and finalised based on the feedback provided by the SMEs.

Advantages

- AcciMap enables the identification of the network of contributory factors underpinning the incident or event in question.
- AcciMap aligns with state-of-the-art accident causation models (e.g., Rasmussen, 1997), and at the time of writing is arguably the most commonly applied accident analysis method (Salmon et al., 2020).
- The AcciMap method is simple to learn and use.
- It is possible to consider contributory factors across the overall sociotechnical system, including decisions and actions at the regulatory, government, and international levels.
- The output offers a detailed and visual description of the incident.
- AcciMap is a generic approach and can be applied in any domain and any context, including to both adverse events and successes.
- Using AcciMap prevents a blame approach and instead engenders a focus on systems failure.
- The output encourages analysts to identify interventions which focuses on broader system interventions as opposed to a 'fixing broken components' approach.
- The ActorMap is a useful modelling tool in its own right, providing a detailed overview of the stakeholders who shared the responsibility for system performance.
- AcciMap is extremely powerful when used to look at multiple incidents to identify trends and recurring systemic issues (McLean et al., 2020; Salmon et al., 2017).

Disadvantages

- AcciMap can be time-consuming to apply particularly for large and complex incidents.
- The identification of relationships between contributory factors is difficult and analysts have been found to perform poorly in reliability and validity studies (e.g., Hulme et al., 2021; Salmon et al., 2023).
- The quality of the analysis produced is entirely dependent upon the quality of the data collected.
- Its graphical output can become complex and hard to decipher when used to analyse large-and complex incidents or events.

Related methods

AcciMap involves the use of various data collection methods such as structured or semi-structured interviews (e.g., the critical decision method) (Klein et al., 1983), direct observation, walkthrough analysis, and documentation review (e.g., incident reports, standard operating procedures). A number of contributory factor classification schemes have been developed to support the use of AcciMap to analyse multiple incident datasets, such as Salmon et al.'s led outdoor activity incident contributory factor classification scheme (Salmon et al., 2017) and Newnam et al.'s patient handling injury contributory factor classification scheme (Newnam et al., 2021). Based on a meta-analysis of previous AcciMap analyses, Salmon et al. (2020) proposed a generic contributory factor classification scheme that could be used in any domain.

The PreventiMap method was developed to work in conjunction with AcciMap and is used to identify and depict the network of interventions that is required to respond to critical safety issues (Goode et al., 2016). PreventiMaps use the same hierarchical structure and levels as ActorMap and AcciMap and show what interventions are required at each level of the system in question to prevent or manage a particular safety issue. The interventions are linked in a network showing how interventions at one level can support those at other levels. For more information on developing PreventiMap, see Salmon et al. (2022).

Approximate training and application times

AcciMap is simple to learn and apply. Application time is dependent on the incidents or events under analysis, and the method can become time-consuming when applied to large-scale complex incidents. For example, an analysis of a single injury incident can be produced during a one–two hour workshop with SMEs. For more complex incidents or events, timescales are expected to be around one–two weeks for data collection and a further week for the initial construction of the AcciMap.

Reliability and validity

Hulme et al. (2021b) assessed the criterion-referenced validity of AcciMap and found acceptable performance for the placement of contributory factors across the AcciMap levels but poor performance when identifying the relationships between contributory factors. Similarly, Salmon et al. (2023) tested the criterion-referenced validity of AcciMap by providing training for 67 healthcare practitioners and asking them to use AcciMap to analyse an adverse medication administration event. Salmon et al. (2022) reported that participants achieved high levels of validity for the placement of contributory factors at the correct healthcare system level but only moderate levels when identifying relationships between contributory factors. Finally, a more recent intra-rater and inter-rater reliability study found similar results, specifically that reliability was higher for the placement of contributory factors across the RMF compared to the modelling of relationships across the hierarchy (Hulme et al., 2023). Accordingly, Salmon et al.'s (2020) AcciMap contributory factor classification scheme can be used to alleviate some of the concerns around reliability and validity.

Tools required

AcciMaps can be developed using pen and paper; however, drawing software packages such as Microsoft Visio, PowerPoint, or Lucidchart are often used to create the final ActorMap and AcciMap diagrams. Qualitative analysis software tools, such as Nvivo, are useful for coding and aggregating findings across multiple data sources and AcciMaps.

Case study examples

To demonstrate the flexibility and scalability of AcciMap, we present three case study applications of AcciMap: (1) an analysis of a single incident, (2) a multiple incident analysis, and (3) an analysis of a successful event.

Single incident example: athlete injury incident

The single incident AcciMap presented below is based on the following hypothetical example injury to an international football player (Figure 11.10).

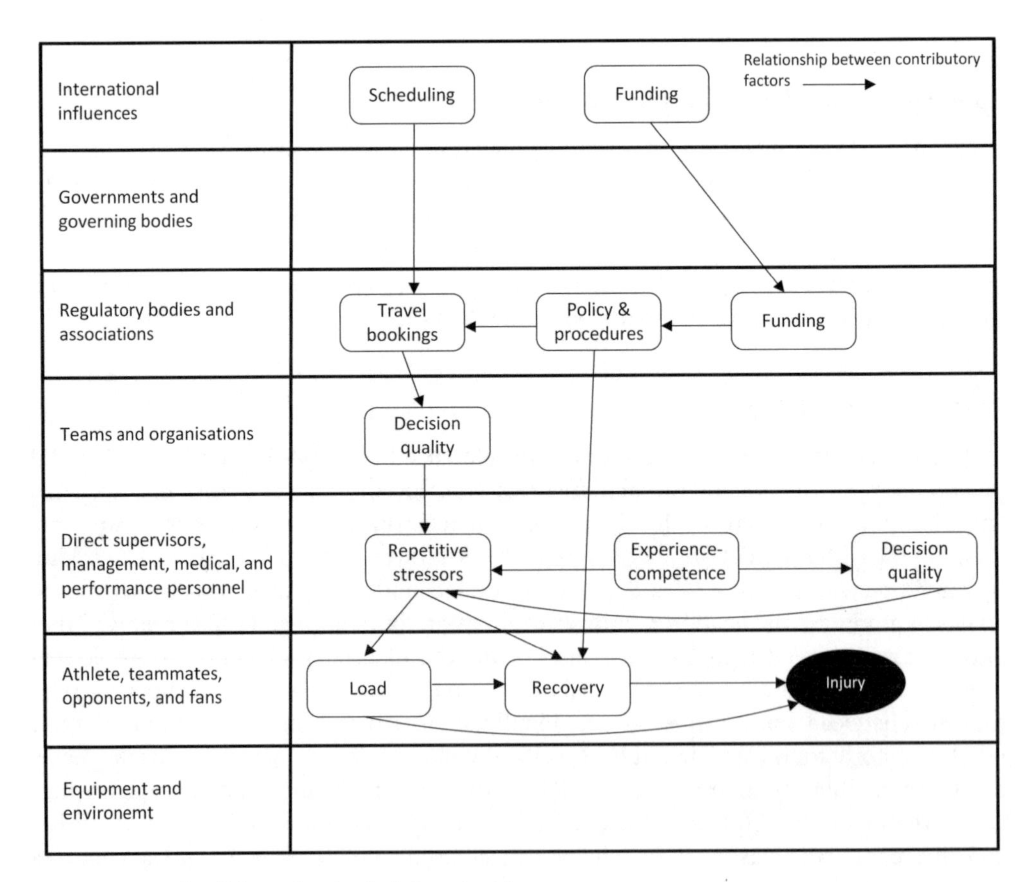

FIGURE 11.10 AcciMap of a single injury incident.

I suffered a hamstring strain in the 75th minute of my first game back after returning to my club in England from two international friendly's back in my home country. It was my 3rd game in 10 days and occurred 2 days after I returned on a long-haul flight. I mentioned to my club physio I was tight after the second international, but the coach insisted I play as other players had returned to the club one day after me, and he thought I was fresher. In the national team camp we only had one physio, as two senior physios were isolating with Covid. I was also forced to travel economy class on a full flight, as my football association is currently having financial issues, and cutting budgets.

Aggregated incident example: Incident causation in led outdoor recreation.

The aggregated incident analysis is based on data derived from the Understanding and Preventing Led Outdoor Accidents Data System (UPLOADS) incident reporting and learning system (Goode et al., 2018; McLean et al., 2020; Salmon et al., 2017). The UPLOADS programme of research involved the development and implementation of an incident reporting and learning system and national incident dataset, underpinned by Rasmussen's RMF and AcciMap method (Salmon et al., 2017). Specifically, the data reported from the Australian Led Outdoor Activity (LOA) sector is used to produce AcciMaps of aggregated incidents for different activities that participants engage in during LOA programmes (e.g., kayaking, hiking, archery, and high ropes courses). The outputs of the UPLOADS allow users to investigate incidents from a systems perspective and identify emergent trends and recurrent issues. This goal is to assist industry associations and government agencies to make evidence-based decisions about issues that affect those involved in the provision of LOA programmes. The AcciMaps presented in Figures 11.11 and 11.12 represent 2311 injury, illness, and psychosocial incidents from the Australian LOA sector between 2020 and 2021. For clarity of the reported contributory factors and relationships, we have presented separate figures. Figure 11.11 shows the contributory factors from across the LOA system, and 11.12 highlights a selection of the relationships between a group of contributory factors.

Successful event: sub-2-hour marathon

To demonstrate the use of AcciMap to analyse successful performance in sport, below we have identified and mapped the enabling factors that contributed to Eluid Kipchoge's assisted sub-2-hour marathon of 1 hr 59 mins, and 40 sec (Figure 11.13). The Ineos 1:59 Challenge in Vienna, Austria was specifically set up for Kipchoge to break the previously unbroken two-hour barrier for the marathon (42km). The challenge did not meet certain criteria for official record eligibility, such as it not being an open event, and Kipchoge was assisted by an alternating team of pacesetters who ran in an aerodynamically favourable formation. Moreover, he was provided support for hydration, and provided a pacing car that projected a laser beam to mark the ideal position on the road. All these mechanisms were implemented to optimise performance and were not in line with the standard rules governing record-eligible marathons. The feat was a testament to how a carefully engineered environment and team support can propel an athlete to achieve what was once thought impossible.

Level	Actor	Contributory factors
Governance, Education & Regulation	State/Fed Government	Communication (3)
	Local Government	Facilities (1)
	Regulators	Standards (1)
Clients	Schools	Communication (17); Compliance (11); Decisions (11); Knowledge (16); Preparation (13); Policies/Procedures (2)
	Parents	Communication (36); Knowledge (12); Preparation (13); Decisions (13)
Planning & Management	Management	Communication (6); Training (8); Supervision (4); Compliance (4); Culture (4); Decisions (10); Budget (1)
		Staffing (6); Knowledge (5); Policies/Procedures (6); Risk Management (7)
	Program	Location (12); Scheduling (11); Suitability (20); Resourcing (4); Supervisor Ratio (2)
People involved in the Incident	Other People	Attitudes (9); Communication (12); Compliance (9); Decisions (21); Experience (9); Knowledge (19); Mental/Physical (5); Preparation (3)
	Group	Composition (8); Peer Interactions (32); Teamwork (15); Group Size (3); Timing (6)
	Supervisors	Attitudes (15); Communication (49); Compliance (14); Decisions (39); Knowledge (23); Mental/Physical (31); Preparation (12)
	Participant	Attitudes (109); Communication (72); Compliance (175); Decisions (399); Experience (392); Knowledge (205); Mental/Physical (624); Preparation (102)
	Leader	Attitudes (7); Communication (107); Compliance (16); Decisions (79); Experience (37); Knowledge (51); Mental/Physica (32); Preparation (23)
Resources & Environment	Environment	Animals/Insects (205); Facilities (95); Terrain (501); Water (33); Weather (202); Trees/Vegetation (172)
	Resources	Food/Drink (96); Medication (14); Documentation (12); Equipment/Clothing (408)

FIGURE 11.11 UPLOADS AcciMap showing the contributory factors involved 2311 reported incidents.

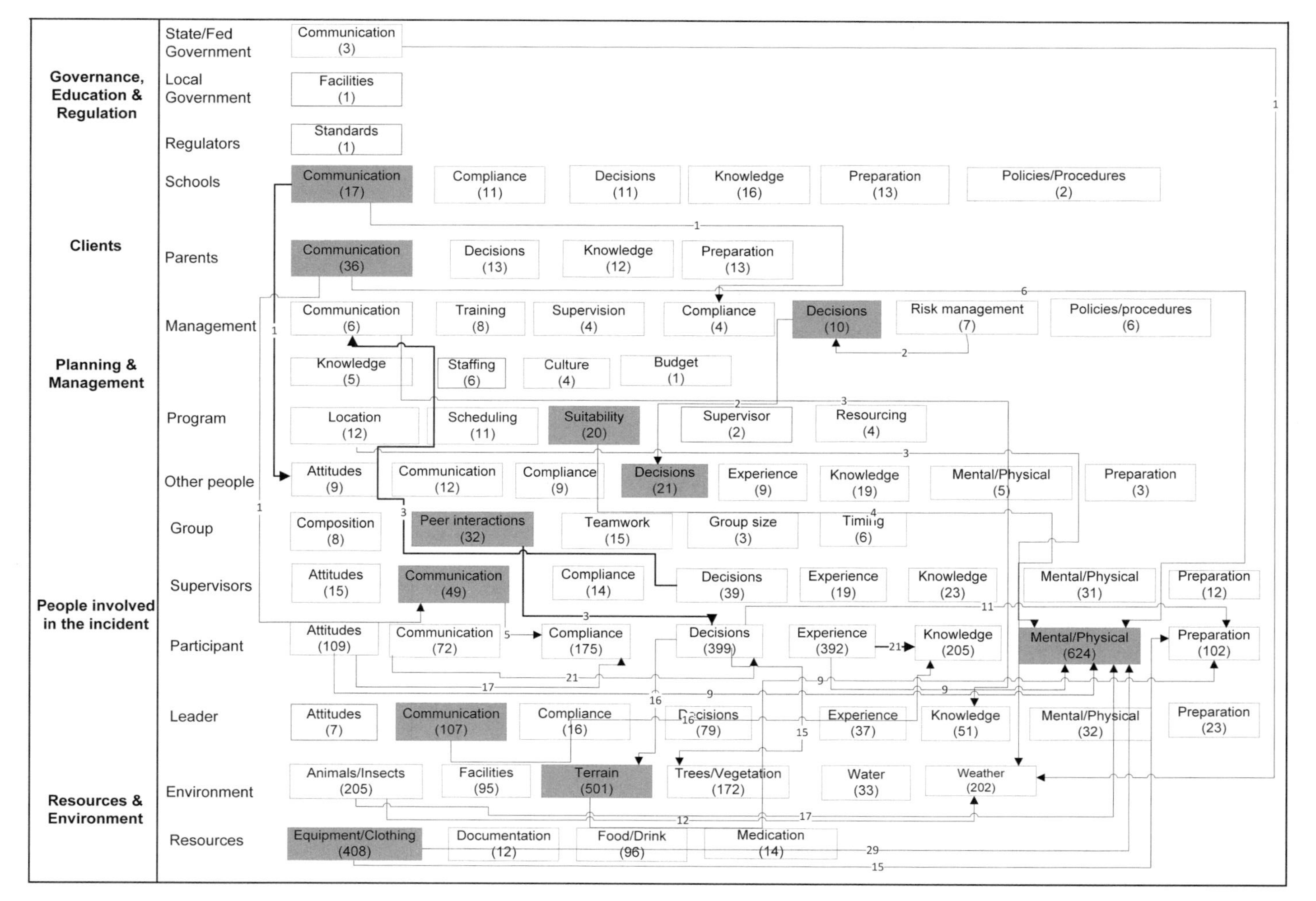

FIGURE 11.12 UPLOADS AcciMap highlighting a selection of nodes and relationships between contributory factors in 2311 reported incidents. For clarity, the lines separating the levels have been removed.

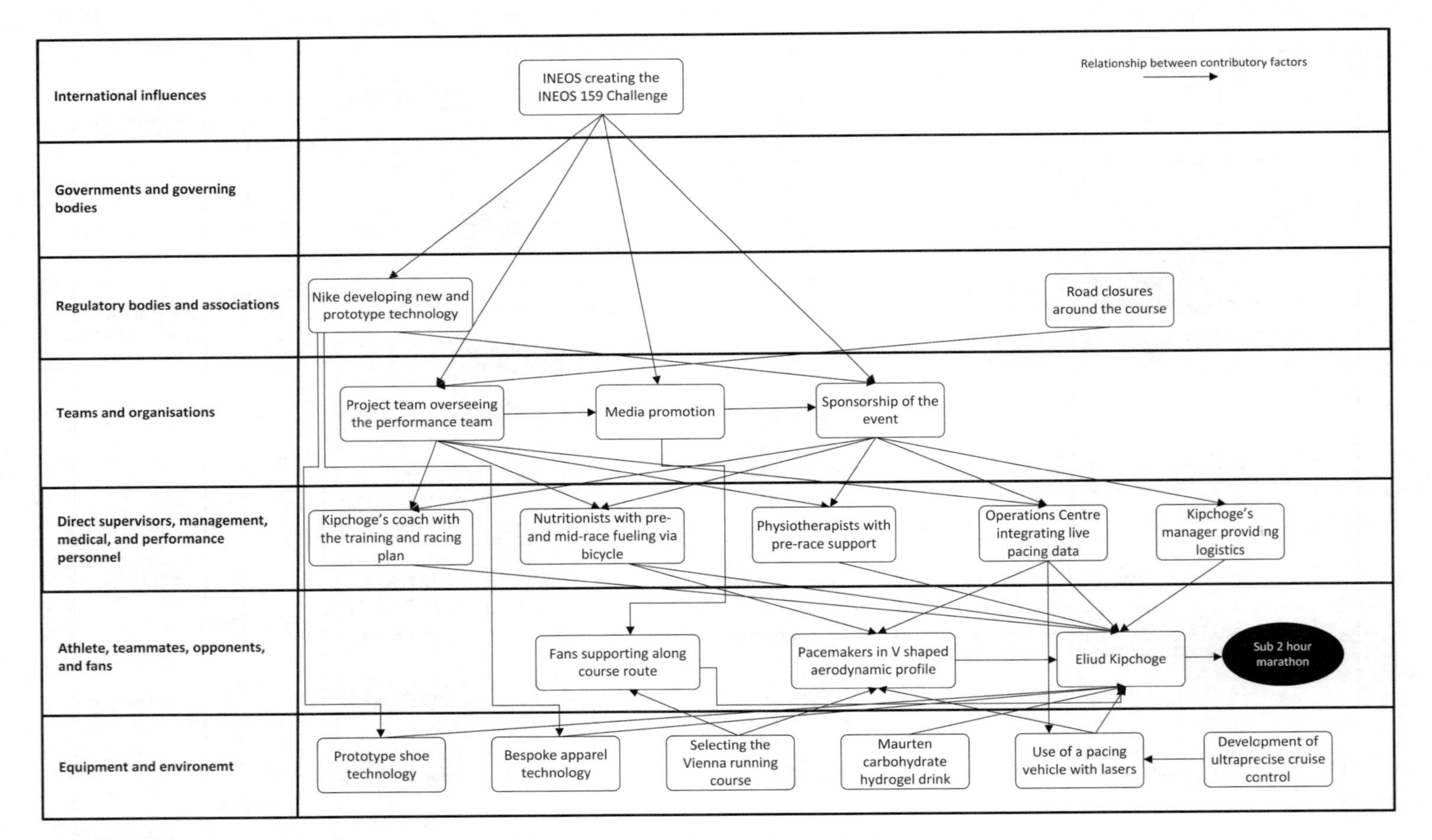

FIGURE 11.13 AcciMap of Eluid Kipchoge's assisted marathon of 1 hr 59 mins, and 40 sec.

Discussion

The three case study applications presented in this chapter have demonstrated the utility of AcciMap to identify and visually represent the contributory factors involved in single adverse events, aggregated incidents, and successful performances.

In the single injury AcciMap, the case study analysis enabled the identification of contributory factors beyond the typical root cause analysis, e.g., the physiological mechanism of injury, shedding light on underlying systemic issues influencing these occurrences (Salmon et al., 2013). In sport, when an athlete is injured or under performs, it's rarely just about that moment or that individual. Often, it is about a combination of factors such as the cumulative weight of training and match schedules, equipment choices, and even organisational culture. The case study example demonstrated that scheduling on matches at an international level, travel bookings at a Football Association level, and decisions and actions of coaches and performance staff each contributed to the injury incident. By mapping out the contributing factors across the systems levels, AcciMap enables sports organisations to pinpoint specific areas of intervention, mitigate risks, and implement informed policies that enhance safety and performance.

Where single incidents provide useful snapshot insights, aggregated incident analysis can reveal causal trends and patterns. Sports organisations often grapple with recurring issues, e.g., hamstring injuries (Ekstrand et al., 2023), persistent equipment failures, or systemic issues such as doping or child safeguarding problems (Dodd et al., 2023; McLean et al., 2023; Naughton et al., 2024). When used with an appropriate classification scheme, AcciMap's structured approach allows for the aggregation of incident data to highlight trends and systemic vulnerabilities, that can inform long-term strategic changes, reshape training programmes, influence equipment design, and even prompt shifts in organisational structures (McLean et al., 2021). By identifying these patterns, sports organisations can transition from reactive to proactive, preventing potential incidents before they occur.

Lastly, sporting success appears to be as complex and layered as sports failures. When a team wins a title or an athlete breaks a world record, it is not merely a moment of individual triumph but the culmination of a series of interacting systemic factors. The AcciMap of the sub-2-hour marathon details the enabling factors from across the system and included sponsors, local governments, support personnel, and a range of technologies. By mapping out the factors that lead to success, organisations can foster a culture of excellence, replicate effective strategies, and build on the positive attributes of their systems.

Recommended Reading

Naughton, M., Salmon, P. M., Kerherve, H., & McLean, S. (2024). Applying a systems thinking lens to anti-doping: A systematic review identifying the contributory factors to doping in sport. *Journal of Sports Sciences*. Advance online publication.

Dodd, K., Solomon, C., Naughton, M., Salmon, P. M., & McLean, S. (2023). What enables child sexual abuse in sport? A systematic review. *Trauma, Violence, & Abuse*, 15248380231190666.

Rasmussen, J. (1997). Risk management in a dynamic society: A modelling problem. *Safety Science*, 27(2/3), 183–213.

Svedung, I., & Rasmussen, J. (2002). Graphic representation of accident scenarios: Mapping system structure and the causation of accidents. *Safety Science*.

Salmon, P. M., Hulme, A., Walker, G. H., Berber, E., Waterson, P., & Stanton, N. A. (2020). The big picture on accident causation: A review, synthesis and meta-analysis of AcciMap studies. *Safety Science*, 126, 1–15.

References

Australian Transport Safety Bureau. (2008). *Analysis, causality and proof in safety investigations. Aviation Research and Analysis Report – AR-2007–053*. Canberra: Australian Transport Safety Bureau.

Branford, K. (2007). An investigation into the validity and reliability of the AcciMap approach. Unpublished doctoral dissertation, Australian National University, Canberra, Australia.

Branford, K., Naikar, N., & Hopkins, A. (2009). Guidelines for AcciMap analysis. In A. Hopkins (Ed.), *Learning from High Reliability Organisations*. Sydney: CCH, 193–212.

Cassano-Piche, A. L., Vicente, K. J., & Jamieson, G. A. (2009). A test of Rasmussen's risk management framework in the food safety domain: BSE in the UK. *Theoretical Issues in Ergonomics Science*, 10(4), 283–304.

Dodd, K., Solomon, C., Naughton, M., Salmon, P. M., & McLean, S. (2024). What enables child sexual abuse in sport? A systematic review. *Trauma, Violence, & Abuse*, 25(2), 1599–1613.

Ekstrand, J., Bengtsson, H., Waldén, M., Davison, M., Khan, K. M., & Hägglund, M. (2023). Hamstring injury rates have increased during recent seasons and now constitute 24% of all injuries in men's professional football: The UEFA Elite Club Injury Study from 2001/02 to 2021/22. *British Journal of Sports Medicine*, 57(5), 292–298.

Goode, N., Salmon, P. M., Lenne, M. G., & Hillard, P. (2014). Systems thinking applied to safety during manual handling tasks in the transport and storage industry. *Accident Analysis & Prevention*, 68, 181–191.

Goode, N., Salmon, P. M., Lenne, M. G., & Finch, C. (2018). *Bridging the gap between accident prevention theory and practice*. Boca Raton, FL: CRC Press.

Goode, N., Read, G. J., van Mulken, M. R., Clacy, A., & Salmon, P. M. (2016). Designing system reforms: Using a systems approach to translate incident analyses into prevention strategies. *Frontiers in Psychology*, 7, 1974.

Goode, N., Salmon, P. M., Taylor, N. Z., Lenné, M. G., & Finch, C. F. (2017). Developing a contributing factor classification scheme for Rasmussen's AcciMap: Reliability and validity evaluation. *Applied Ergonomics*, 64, 14–26.

Hulme, A., Stanton, N. A., Waterson, P., Walker, G. H., & Salmon, P. M. (2019). What do applications of systems thinking accident analysis methods tell us about accident causation? A systematic review of applications between 1990 and 2018. *Safety Science*, 117, 164–183.

Hulme, A., Stanton, N. A., Walker, G. H., Waterson, P., & Salmon, P. M. (2021a). Complexity theory in accident causation: Using AcciMap to identify the systems thinking tenets in 11 catastrophes. *Ergonomics*, 64(7), 821–838.

Hulme, A., Stanton, N. A., Walker, G. H., Waterson, P., & Salmon, P. M. (2021b). Are accident analysis methods fit for purpose? Testing the criterion-referenced validity of AcciMap, STAMP-CAST and AcciNet. *Safety Science*. Accepted for publication 16th August 2021.

Hulme, A., Stanton, N. A., Walker, G. H., Waterson, P., & Salmon, P. M. (2023). Testing the reliability of accident analysis methods: A comparison of AcciMap, STAMP-CAST and AcciNet. *Ergonomics*, 67(5), 695–715.

Jenkins, D. P., Salmon, P. M., Stanton, N. A., & Walker, G. H. (2010). A systemic approach to accident analysis: A case study of the Stockwell shooting. *Ergonomics*, 53, 1–17.

Johnson, C. W., & de Almeida, I. M. (2008). An investigation into the loss of the Brazilian space programme's launch vehicle VLS-1 V03. *Safety Science*, 46(1), 38–53.

Lane, B., Read, G. J. M., & Salmon, P. M. (2020). A systems thinking perspective on the barriers to treatment access for people with eating disorders. *International Journal of Eating Disorders*, 53(2), 174–179.

Le Coze, J. C. (2010). Accident in a French dynamite factory: An example of an organisational investigation. *Safety Science*, 48(1), 80–90.

Le Coze, J. C. (2015). Reflecting on Jens Rasmussen's legacy. A strong program for a hard problem. *Safety Science*, 71, 123–141.

Lintern, G. (2019). Jens Rasmussen's risk management framework. *Theoretical Issues in Ergonomics Science*, 21(1), 56–88.
McLean, S., Finch, C. F., Goode, N., Clacy, A., Coventon, L. J., & Salmon, P. M. (2021). Applying a systems thinking lens to injury causation in the outdoors: Evidence collected during 3 years of the understanding and preventing led outdoor accidents data system. *Injury Prevention*, 27(1), 48–54.
McLean, S., Naughton, M., Kerhervé, H., & Salmon, P. M. (2023). From anti-doping-I to anti-doping-II: Toward a paradigm shift for doping prevention in sport. *International Journal of Drug Policy*, 115, 104019.
Naughton, M., Salmon, P. M., Kerherve, H., & McLean, S. (2024). Applying a systems thinking lens to anti-doping: A systematic review identifying the contributory factors to doping in sport. *Journal of Sports Sciences*. Advance online publication, 1–15.
Newnam, S., & Goode, N. (2015). Do not blame the driver: A systems analysis of the causes of road freight crashes. *Accident Analysis & Prevention*, 76, 141–151.
Newnam, S., Goode, N., Salmon, P., & Stevenson, M. (2017). Reforming the road freight transportation system using systems thinking: An investigation of Coronial inquests in Australia. *Accident Analysis & Prevention*, 101, 28–36.
Newnam, S., Goode, N., Read, G. J., Salmon, P. M., & Gembarovski, A. (2021). Systems-thinking in action: Results from implementation and evaluation of the patient handling injuries review of systems Toolkit. *Safety Science*, 134, 105086.
Rasmussen, J. (1997). Risk management in a dynamic society: A modelling problem. *Safety Science*, 27(2/3), 183–213.
Salmon, P. M., King, B., Hulme, A., Chari, S., McCormack, L., Tresillian, M.,... & Read, G. J. (2023). Toward the translation of systems thinking methods in patient safety practice: Assessing the validity of Net-HARMS and AcciMap. *Safety Science*, 159, 106003.
Salmon, P. M., Coventon, L., & Read, G. J. (2022). A systems analysis of work-related violence in hospitals: Stakeholders, contributory factors, and leverage points. *Safety Science*, 156, 105899.
Salmon, P. M., Cornelissen, M., & Trotter, M. (2012). Systems-based accident analysis methods: A comparison of Accimap, HFACS, and STAMP. *Safety Science*, 50(4), 1158–1170.
Salmon, P. M., Read, G. J., Stanton, N. A., & Lenné, M. G. (2013). The crash at Kerang: Investigating systemic and psychological factors leading to unintentional non-compliance at rail level crossings. *Accident Analysis & Prevention*, 50, 1278–1288.
Salmon, P. M., Coventon, L., & Read, G. J. M. (2021). Understanding and preventing work-related violence in hospital settings: A systems thinking approach. *SafeWork New South Wales Final Report*, https://www.safework.nsw.gov.au/__data/assets/pdf_file/0006/964716/understanding-and-preventing-work-related-violence-in-hospital-settings-a-systems-thinking-approach.pdf
Salmon, P. M., Goode, N., Lenné, M. G., Finch, C. F., & Cassell, E. (2014a). Injury causation in the great outdoors: A systems analysis of led outdoor activity injury incidents. *Accident Analysis & Prevention*, 63, 111–120.
Salmon, P. M., Goode, N., Archer, F., Spencer, C., McArdle, D., & McClure, R. J. (2014b). A systems approach to examining disaster response: Using Accimap to describe the factors influencing bushfire response. *Safety Science*, 70, 114–122.
Salmon, P. M., Goode, N., Taylor, N., Dallat, C., Finch, C., & Lenne, M. G. (2017). Rasmussen's legacy in the great outdoors: A new incident reporting and learning system for led outdoor activities. *Applied Ergonomics*. 59, Part B, 637–648.
Salmon, P. M., Hulme, A., Walker, G. H., Berber, E., Waterson, P., & Stanton, N. A. (2020). The big picture on accident causation: A review, synthesis and meta-analysis of AcciMap studies. *Safety Science*, 126, 1–15.
Svedung, I., & Rasmussen, J. (2002). Graphic representation of accident scenarios: Mapping system structure and the causation of accidents. *Safety Science*, 40(5), 397–417.
Thoroman, B., Goode, N., Salmon, P., & Wooley, M. (2019). What went right? An analysis of the protective factors in aviation near misses. *Ergonomics*, 62(2), 192–203.

Trotter, M. J., Salmon, P. M., & Lenne, M. G. (2014). Impromaps: Applying Rasmussen's risk management framework to improvisation incidents. *Safety Science*, 64, 60–70.

Vicente, K. J., & Christoffersen, K. (2006). The Walkerton E. coli outbreak: A test of Rasmussen's framework for risk management in a dynamic society. *Theoretical Issues in Ergonomics Science*, 7(02), 93–112.

Waterson, P. (2009). A systems ergonomics analysis of the Maidstone and Tunbridge wells infection outbreaks. *Ergonomics*, 52(10), 1196–1205.

Waterson, P., Jenkins, D. P., Salmon, P. M., & Underwood, P. (2017). 'Remixing Rasmussen': The evolution of Accimaps within systemic accident analysis. *Applied Ergonomics*, 59, 483–503.

Waterson, P. E., & Jenkins, D. P. (2010). Methodological considerations in using AcciMaps and the risk management framework to analyse large-scale systemic failures. In *5th IET International Conference on System Safety 2010* (pp. 1–6). IET.

Waterson, P. E., & Jenkins, D. T. (2011), Lessons learnt from using Accimaps and the risk management framework to analyse large-scale systemic failures. In M. Anderson (Ed.) *Contemporary ergonomics* (pp. 6–13). London: Taylor & Francis.

12
THE ACCIDENT NETWORK (ACCINET) METHOD

Background

The Accident Network (AcciNet) method (Salmon et al., 2020) is a systems theory-based accident analysis method that was developed to address some of the known limitations associated with existing incident analyses methods (Salmon et al., 2022). AcciNet uses a task network and associated classification scheme to identify the network of tasks and contributory factors, including normal performance, involved in accident events. Specifically, AcciNet uses the same task network as the Networked Hazard Analysis and Risk Management System (Net-HARMS) method (Chapter 10) to identify contributory factors and describe incident aetiology within complex systems. From a theoretical standpoint, AcciNet is based on the following fundamental tenets of accident causation (Dekker, 2011; Leveson, 2004; Rasmussen, 1997):

1. That all accidents are created by an interacting network of behaviours associated with multiple actors, both human and non-human, who reside across the sociotechnical system of interest (Rasmussen, 1997). In a sports context, this means that injury or poor performances are created by the decisions and actions of multiple stakeholders, not just by sub-standard athlete or coach decisions and actions;
2. That the interacting network of contributory factors involved in accidents includes 'work as imagined' (i.e., undertaken in line with procedures), 'normal performance' (i.e., 'work as done' where performance was appropriate and no discernible failure occurred), and decisions and actions whereby performance can reasonably be classified as sub-optimal (Dekker, 2011; Hollnagel, 2012). In a sports context, this means that injury or poor performances are created by the interactions between appropriate decisions and actions and sub-optimal decisions and actions; and
3. That 'emergent risks' play a critical role in accident causation. Emergent risks occur when multiple behaviours interact with one another to create unexpected and difficult to foresee outomes. These emergent risks occur across the sociotechnical system and

DOI: 10.4324/9781003259473-15

may be proximal or distal to the accident event. An example of emergent risk in a sports context would be an inappropriate training load imposed on athletes that arises from the interaction between a congested match schedule, performance pressures, and an inexperienced coach.

AcciNet was developed to support analysts in identifying and depicting the three tenets of accident causation described above. The method uses a task network and actor network (ActorNet) for the system under analysis as its primary input, with analysts subsequently using both networks to identify contributory factors, their interrelations, and associated actors. A classification scheme (Dallat et al., 2018; Salmon et al., 2020) can then be used to classify contributory factors further if needed; however, this is included only as an optional step if required as part of the investigation or analysis aims.

AcciNet outputs include a task network showing the interaction between tasks and contributory factors and a table describing each contributory factor and potential safety interventions designed to prevent similar future occurrences. A key feature of the AcciNet method is the use of the task network which removes the need for analysts to identify the relationships between contributory factors (as the relationships between tasks are already specified within the task network). A final notable feature of AcciNet is it was developed to be applied in conjunction with the Net-HARMS (Dallat et al., 2018) risk assessment method (see Chapter 10) whereby the same task network supports both risk assessment and accident analysis for the system in question. This enables analysts and practitioners to integrate their prospective risk assessment and retrospective incident analysis activities. For example, injury incident analyses based on AcciNet can be fed into prospective Net-HARMS risk assessments either to add to the risks identified or to provide an indication of where identified risks have occurred previously. Moreover, risk controls that have failed (as identified during the AcciNet analysis) can be noted during risk assessment activities. During AcciNet analyses the relevant risk assessment outputs, based on the same task network, can be used to support the identification of incident contributory factors. Specifically, analysts can first review the risks identified and then seek to determine whether they played a role in the injury incident. This also enables the accuracy of risk assessments to be verified as well as the effectiveness of risk controls.

Applications in sport

AcciNet is a generic approach that can be applied in all safety-critical domains, including within sport to analyse, injury incidents, recurring issues, and performance failures. AcciNet will be particularly useful in sport when used in conjunction with Net-HARMS, with both analyses based on the same task network. This will allow a prospective assessment of potential risks followed by an in-depth analysis of adverse events as they occur.

Procedure and advice

A flowchart depicting the AcciNet procedure is presented in Figure 12.1. Step-by-step guidance is presented below.

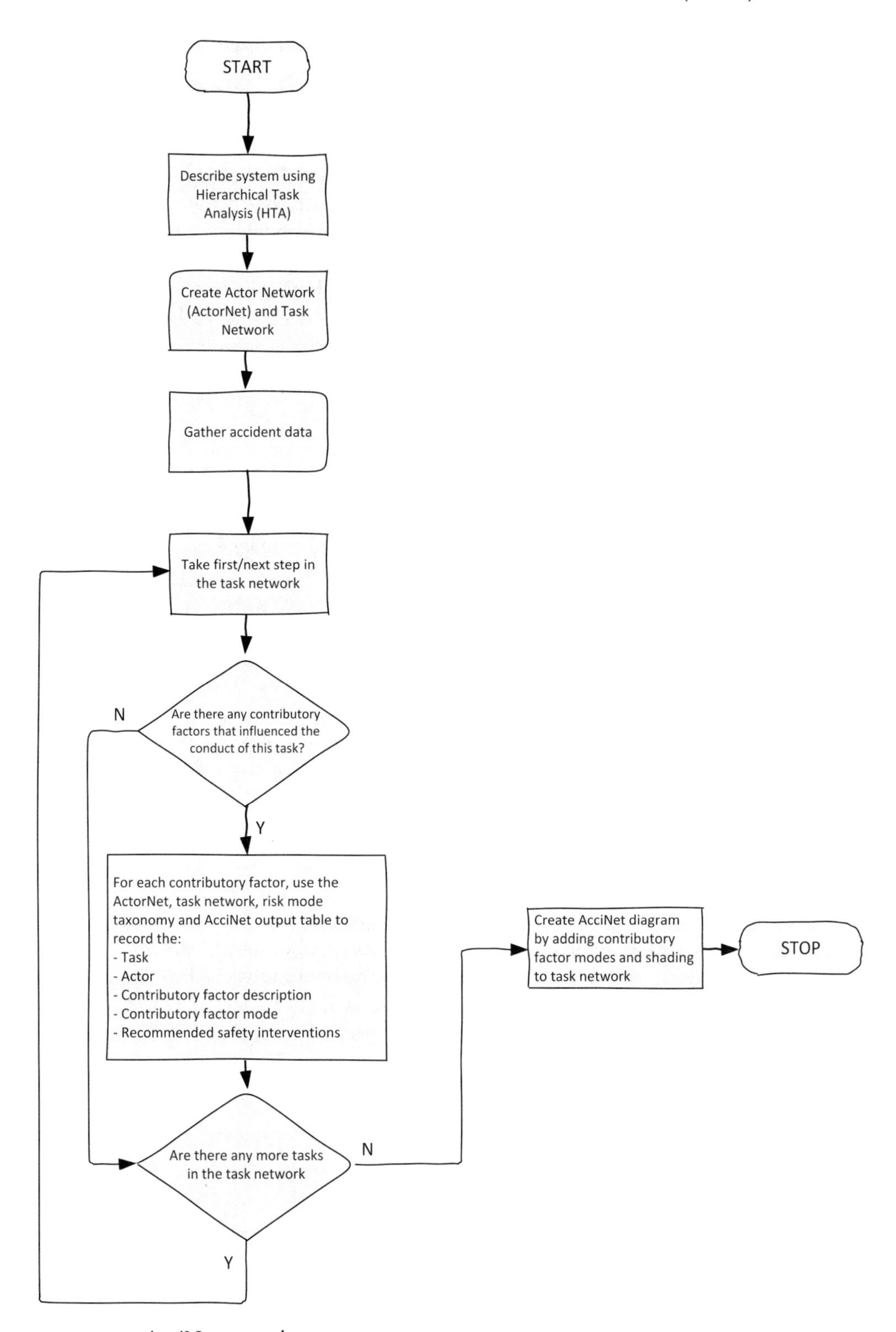

FIGURE 12.1 AcciNet procedure.

Step 1: Define aims of the analysis and the incident and system under analysis

The initial step involved in applying AcciNet is to clearly define the aims of the analysis, the incident or issue to be analysed, and any analysis boundaries. For example, in the case study presented in this chapter, the aim was to examine the rapid downfall of Leeds United Football Club during the late 1990s and early 2000s to identify the contributory factors that interacted to create the incident. While applications of AcciNet will always involve the analysis of an incident or set of incidents that form a recurring issue, the aims may differ in terms of whether the intention is to understand the contributory factors involved or whether it is to develop safety interventions designed to prevent similar incidents from occurring in future. Defining the analysis boundary is important as it dictates how far reaching the analysis will be, how far back in time the analysis will go, and what data collection methods are most appropriate.

Step 2: Construct a hierarchical task analysis for the system under analysis

Once the incident under analysis and aims of the analysis are clearly defined, a hierarchical task analysis (HTA) (see Chapter 3) for the system under analysis should be created. Initially, this involves collecting data regarding the goals, tasks, and operations undertaken within the system under analysis. Data should be collected regarding the goals and tasks involved, the human and non-human agents involved, the interactions between humans and non-human agents, the ordering of tasks and information on the factors that influence behaviour, task performance, and task outcome. A number of different approaches can be used to collect this data, including observations, concurrent verbal protocols, structured or semi-structured interviews, e.g., the Critical Decision Method (CDM) (Klein et al., 1989), questionnaires and surveys, walkthrough analysis and documentation review (e.g., incident reports, standard operating procedures). The data collection approach selected is dependent upon project constraints, such as time, resources, access, and the number of analysts available.

A key focus of any AcciNet analysis should be on tasks that are undertaken across the system, including supervisory, management, regulatory, and government activities. Developing the HTA itself involves identifying the main goal of the system under analysis and then decomposing this into a series of sub-goals, operations, and plans. Once the initial draft HTA is complete it is useful to have various SMEs review it. The HTA should then be refined based on SME feedback. It is normal practice for the HTA to go through multiple iterations before it is finalised.

Step 3: Create task network

Once the HTA is finalised, the next step involves constructing a task network to support the AcciNet analysis. Task networks (see Figure 12.2) are used to represent HTA outputs in the form of a network which shows key system tasks and the relationships between them (Stanton et al., 2013; 2018). This enables analysts to understand the interactions and interrelations between tasks across the work system. Within Figure 12.2 the circular nodes represent tasks and the arrows linking the tasks represent relationships between

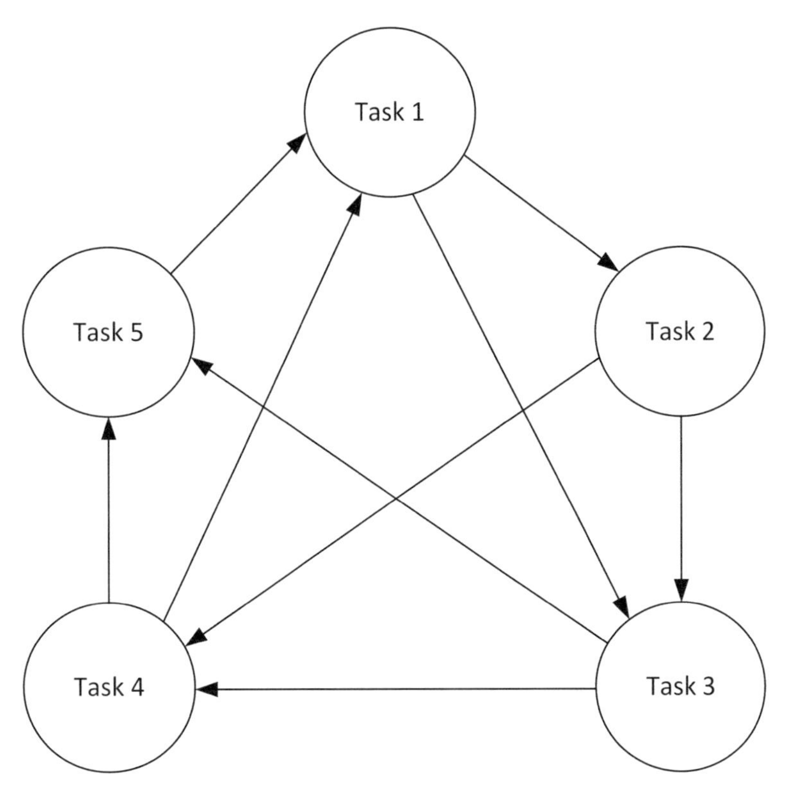

FIGURE 12.2 Task network example showing the relationships between five tasks.

tasks. Relationships between tasks are included in the task network when (Salmon et al., 2022):

1 Tasks are undertaken sequentially, e.g., task three is undertaken after completion of task two;
2 Task are undertaken together, e.g., tasks one and two are undertaken together;
3 The outcomes of one task influence the conduct of another, e.g., the outcomes of task four influence how task five is undertaken; or
4 The conduct of one task is dependent on completion of the other, e.g., task five cannot be undertaken until task four is completed.

Task networks are constructed by taking the first layer of sub-goals from the HTA and identifying which of them are related with one another based on relationship types described above. For example, in the Leeds United case study presented at the end of this chapter, the tasks 'Financial management' and 'Player and staff recruitment and retention' are linked because the club's financial management involves setting and managing the club's budget, which includes various expenses such as player wages, transfer fees, and other operational costs (Figure 12.3).

Once a draft task network is developed it should be reviewed by relevant SMEs and refined based on their feedback. This involves asking them to review and verify the tasks,

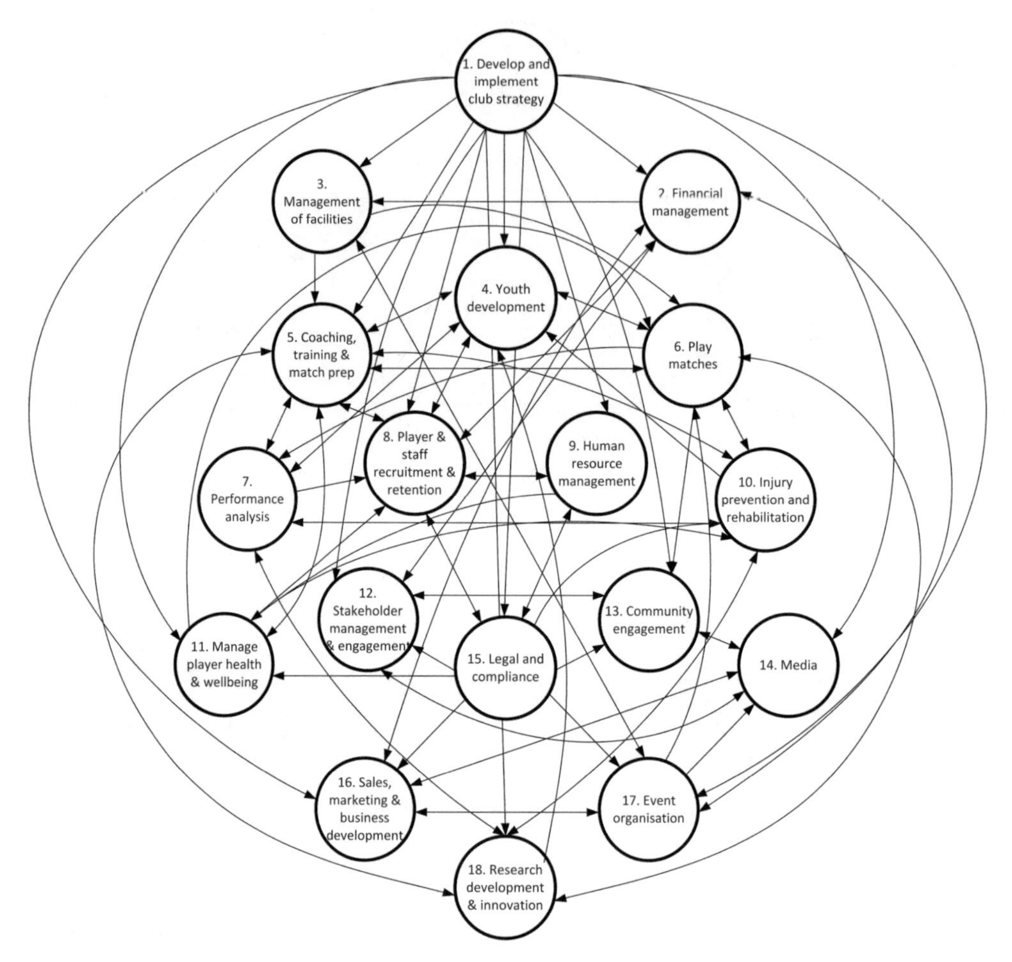

FIGURE 12.3 'English Premier League football club operation' task network. Relationships in the task network represent instances where the conduct of one task influences, is undertaken in combination with, or is dependent on, another task.

the relationships between the tasks, and the terminology used throughout the task network. Changes at this stage typically involve the addition and/or removal of tasks or the relationships between tasks, and the modification of task descriptions. Like HTA, it is normal practice for the task network to go through various iterations before it is finalised.

While we recommend that the development of a task network should be based on HTA, it is possible to develop a task network based directly on raw data, SME input, and document reviews (Stanton, 2014). This approach is typically adopted where project constraints do not enable development of an initial HTA, such as a lack of resources or limited access to data. If a HTA is not developed, we recommend that analysts use SME review to ensure that the core tasks of interest are included in the task network.

Step 4: Gather data regarding incident under analysis

AcciNet is dependent upon the provision of detailed and accurate data regarding the incident or recurring issue under analysis. The next step therefore involves collecting data regarding the incident in question as well as the system in which the incident occurred. Three broad forms of data are required:

- Data on the work activities or processes which occur across the system in which the incident occurred;
- Data on the incident itself and any contributory factors that played a role in its occurrence; and
- Data on the system in terms of who resides within it and shares the responsibility for safe completion of the activities or processes relating to the incident.

Data collection for AcciNet can involve a range of activities, including interviews with those involved in the incident or SMEs for the domain, activity, or type of incident under analysis, reviewing reports or inquiries into the incident, and observing recordings of the incident. It is important during data collection to consider all relevant stakeholders in the system. For example, if using interviews as the primary data collection approach it is useful to interview as many relevant stakeholders as possible, such as athletes, coaches, high-performance staff, club managers and executives, members of the relevant governing body, and so on.

Step 5: Review and refine task network

Once sufficient data regarding the incident and system in question is collected, the next step involves reviewing and refining the draft task network. This involves ensuring that the existing tasks and relationships are appropriate and also adding any new tasks and relationships identified during data collection activities.

Step 6: Develop ActorNet (Optional)

The ActorNet is used to show the different stakeholders who share responsibility for the incident in question and specifically which tasks within the task network they undertake. Development of the ActorNet involves identifying all of the relevant 'actors' who contribute to the tasks and then adding them to the task network based on the tasks that they undertake or contribute to. In the Leeds United case study example actors include:

- Players;
- Coaches and performance staff;
- Chairman; and
- Board of Directors

Step 7: Review tasks and identify relevant contributory factors

Next the data gathered during step 4 should be reviewed in conjunction with the task network to identify relevant contributory factors that played a role in creating the incident under analysis. We define contributory factors as a factor

> that, if it had not occurred or existed at the relevant time, then either the occurrence would probably not have occurred, adverse consequences associated with the occurrence would probably not have occurred or have been as serious, or another contributing safety factor would probably not have occurred or existed.
>
> *(Australian Transport Safety Bureau, 2008)*

It is recommended that analysts take as broad a view as possible when identifying contributory factors. A key requirement here is to look for contributory factors across tasks within the task network and to look for contributory factors associated with the actors identified in the ActorNet. This involves taking each task in the task network and asking whether the conduct of the task or its outcomes played a role in creating the incident. If the answer is yes, then the analyst should then identify the actors associated with the task and document the contributory factor in a table which includes a description of each task and any associated actors and contributory factors. The AcciNet output table (see Table 12.1) develops incrementally as the analysis process unfolds (i.e., at this stage we are only up to column three – the classification scheme and safety interventions will be added in the subsequent phases below). It is recommended that a group modelling process be used to identify contributory factor, using multiple SMEs and practitioners from the organisation/system.

Step 8: Classify contributory factors

Next, all identified contributory factors should be classified using an appropriate contributory factor classification scheme. While early versions of the method used a taxonomy from the Net-HARMS method (Dallat et al., 2018) (see Table 12.2), Salmon et al.'s (2020) AcciMap classification scheme (Table 12.3) can also be used. The process involves taking each contributory factor identified during Step 7 and classifying into a relevant contributory factor type from the classification scheme. It is normally useful to use multiple analysts for this process and discuss each contributory factor until consensus is achieved. In cases where only one analyst is used, it is useful to have a second analyst conduct a portion of the classification and compare levels of agreement between the two analysts.

TABLE 12.1 AcciNet output table

Task	*Actor(s)*	*Contributory factor description*	*Contributory factor classification*	*Recommended safety intervention*
Insert from task network	Insert from ActorNet	Identify from data	See step 7: use the risk modes taxonomy or classification scheme	For example, SME group decision-making; Delphi approach

TABLE 12.2 Net-HARMS risk mode taxonomy

Classification	*Risk Mode*	*Description*
Task	T1 – Task mistimed	Task is undertaken too early or too late within process
	T2 – Task omitted	Task is not undertaken
	T3 – Task completed inadequately	Task is undertaken but is completed in an inadequate manner
	T4 – Inadequate task object	The object(s) used to complete the task are inadequate
	T5 – Inappropriate task	An inappropriate task is performed instead of the required task
Communication	C1 – Information not communicated	Information required to complete the task is not communicated
	C2 – Wrong information communicated	The wrong information is communicated
	C3 – Inadequate information communicated	Information is communicated but is inadequate, e.g., incomplete communication with missing information
	C4 – Communication mistimed	Communication is undertaken too early or too late within process
Environment	E1 – Adverse environmental conditions	Adverse environmental conditions influence task performance

TABLE 12.3 Generic AcciNet Classification Scheme (adapted from Salmon et al., 2020)

Equipment, environment, and surroundings	*Local area government, planning and budgeting and company management*
1 Animal, plant & biological hazards	44 Communication & coordination
2 Built environment & infrastructure	45 Compliance with procedures, violations & unsafe acts
3 Equipment, technology & resources	46 Culture
4 Information & data	47 Financial pressures
5 Noise & visibility	48 Judgement & decision-making
6 Other	49 Other
7 Physical & natural environment	50 Personnel management & recruitment
8 Time-related	51 Planning & preparation
9 Weather & climate	52 Policy & procedures
10 Work environment	53 Qualification, training, experience & competence
Physical processes and actor activities	54 Risk assessment & management
11 Accident event	55 Supervision
12 Activity, work & operations	56 Time-related
13 Adverse events	**Regulatory bodies and associations**
14 Communication & coordination	57 Audits & inspections
15 Compliance with procedures, violations & unsafe acts	58 Communication & coordination
16 Delayed discovery & response	59 Compliance with procedures, violations & unsafe acts

(*Continued*)

TABLE 12.3 (Continued)

Equipment, environment, and surroundings	*Local area government, planning and budgeting and company management*
17 Equipment, technology & environment	60 Culture
18 Group & teamwork	61 Financial pressures
19 Judgement & decision-making	62 Judgement & decision-making
20 Other	63 Planning & preparation
21 Personnel management & workloads	64 Qualification, training, experience & competence
22 Physical & mental condition	65 Regulatory structures & services
23 Planning & preparation	66 Risk assessment & management
24 Qualification, training, experience & competence	67 Standards, policy & regulations
25 Risk assessment & management	68 Time-related
26 Situation awareness	69 Unclear roles & responsibilities
27 Supervision & leadership	**Government policy and budgeting**
28 Time-related	70 Action omitted & failure to act
29 Weather, climate & natural processes	71 Budget & finance
Technical and operational management	72 Communication & coordination
30 Communication & coordination	73 Culture
31 Compliance with procedures, violations & unsafe acts	74 Judgement & decision-making
32 Culture	75 Policy, legislation & regulation
33 Equipment & environmental design	76 Political structures & services
34 Financial pressures	77 Priorities
35 Judgement & decision-making	78 Qualification, training, experience & competence
36 Other	79 Supervision & enforcement
37 Personnel management & recruitment	
38 Planning & preparation	
39 Policy & procedures	
40 Qualification, training, experience & competence	
41 Risk assessment & management	
42 Supervision	
43 Time-related	

Step 9: Overlay contributory factors onto task network

The next step involves overlaying all identified contributory factors onto the task network. This involves adding contributory factors and actors to the relevant tasks in the task network. For example, in the Leeds United case study example, the contributory factor of 'Club developed and implemented high-risk strategy that involved high borrowing on the basis of continual qualification to the Champions League' was added to the task 'Develop and implement club strategy'. If a classification scheme has been used, the relevant code should be added to the task network. For example, for the contributory factor above the code 'T3 – Task completed inadequately' from Table 12.2 would be added, as the task of developing and implementing club strategy was undertaken in a substandard manner. Contributory factors and their associated actor should be shaded red. Tasks within the AcciNet which represent 'work as imagined' or 'normal performance' are then shaded green to show that no failure or inappropriate behaviours occurred. This shading is critical as it shows the interaction between normal and sub-optimal performance.

Step 10: SME review

It is important to have appropriate SMEs review the final AcciNet diagram and output table. It is best practice to use SMEs who were either involved in the incident or who have extensive knowledge of the system and activities in question. The AcciNet and contributory factors table should be updated and finalised based on the feedback provided by the SMEs.

Step 11: Identify appropriate safety interventions

The final step in the AcciNet process involves developing appropriate safety interventions designed to present similar future occurrences. It is recommended that a group involving the original analysts and appropriate SMEs discuss each contributory factor and identify relevant interventions to prevent future occurrences. The interventions identified should be added to the contributory factors table as shown in Table 12.4. Initial testing of the proposed interventions can then be undertaken by adding them to the task network, linking the nodes to existing system tasks, and considering any potential positive or negative knock-on effects or emergent properties. This supports the identification of ways in which a new risk control or intervention may enhance or degrade existing system tasks. Potential interventions can then subsequently be refined based on the findings and tested further until analysts are satisfied that they will have the desired effect.

TABLE 12.4 'Downfall of Leeds United' AcciNet contributory factors output table extract showing example contributory factors across the task network

Task	*Actor(s)*	*Contributory factor description*	*Contributory factor classification*	*Recommended safety intervention*
1 Develop and implement club strategy	Chairman Board of Directors	Club developed and implemented high-risk strategy that involved high borrowing on the basis of continual qualification to the Champions League	T3 – Task completed inadequately	– Financial controls and regulation – Undertake independent risk assessment on new club strategy and policy – Improved strategy development process, including independent review and auditing

Advantages

- AcciNet considers contributory factors across the overall system, providing an exhaustive analysis of adverse incidents.
- AcciNet provides a clear visual representation of the tasks required in a given system as well as the contributory factors that created the incident in question.
- AcciNet is a generic approach which can be applied in any domain.
- The systems perspective offered ensures that there is a focus on systematic improvements rather than on blaming individuals.
- Analysts are not required to describe the relationships between contributory factors as the task network shows the relationships between tasks by default. The identification of relationships between contributory factors is difficult and analysts have been found to perform poorly in AcciMap reliability and validity studies (e.g., Hulme et al., 2021; Salmon et al., 2023).
- AcciNet can be used in conjunction with a taxonomy and/or classification scheme which supports the aggregation of analyses across multiple incidents.
- AcciNet does not focus exclusively on failure, considering both normal and abnormal performance and how both interact to create adverse events.
- AcciNet was designed to be used with Net-HARMS (Chapter 10) as part of an integrated safety management framework.

Disadvantages

- AcciNet can be time-consuming to apply particularly for large and complex incidents.
- The quality of the analysis produced is entirely dependent upon the quality of the data collected.
- AcciNet does not provide a method to identify and develop corrective measures; these are based on the judgement of the analyst.
- The graphical AcciNet output can become complex and hard to decipher when used to analyse large and complex incidents.
- Developing the initial task network requires significant work and can be difficult for novice analysts.
- As a recently introduced concept in safety science, 'normal performance' requires some introduction and may be difficult to grasp for novice analysts.

Related methods

AcciNet uses a task network which is developed based on a HTA (see Chapter 3) of the system under analysis. AcciNet was designed to be used in conjunction with Net-HARMS (Chapter 10) and SafetyNet as part of the Systems Thinking Accident and Risk Toolkit (START) (Salmon et al., 2022). The integrated toolkit is driven by two components that are shared across the three methods: the task network and the risk mode classification scheme. The task network provides the description of activity that is the basis on which to identify risks during a prospective risk assessment (Net-HARMS), to identify contributory factors during accident analysis (AcciNet), and to test and refine new controls or safety interventions developed based on risk assessment or accident analysis outputs (SafetyNet).

The task network thus ensures that organisational risk assessments, accident analyses, and the evaluation of new risk controls and safety interventions are undertaken based on the same description of the work system.

Approximate training and application times

AcciNet is a relatively simple method to learn and apply but does involve multiple steps of varying duration. For example, data collection and development of the task network can be time-consuming, especially for complex systems involving multiple interrelated tasks. When analysing complex systems and incidents, the application time can run into weeks or even months; however, for simpler systems and incidents it is possible to complete AcciNet analyses in a few hours.

Reliability and validity

AcciNet was only recently introduced (Salmon et al., 2020b) and hence there is little reliability and validity evidence associated with the method. Hulme et al. (2021b) tested the criterion-referenced concurrent validity of AcciNet, AcciMap (Chapter 11), and STAMP-CAST when used by 34 Human Factors and safety practitioners to analyse an automated vehicle collision. The findings revealed that AcciNet achieved a moderate positive correlation coefficient against an expert 'gold standard' incident analysis (Hulme et al., 2021b). This is promising and means that AcciNet achieved respectable levels of validity overall when assigning and classifying contributory factors to tasks within the task network.

Tools needed

AcciNet can be conducted using pen and paper; however, the AcciNet output table is normally undertaken using a Microsoft Excel spreadsheet or Microsoft Word document with the table function selected. Task networks are typically drawn in Microsoft Visio or a related drawing programme.

Case study application: downfall of Leeds United Football Club

On the opening day of the 2007/2008 English football season Leeds United Football Club lost to opponents Tranmere Rovers FC. In less than a decade, gross financial mismanagement saw the club go from challenging for the Premier League title and European Champions League to playing in the third tier of English football for the first time in their history. This failure is now referred to in pop culture as 'doing a Leeds' and was the product of multiple interacting contributory factors. This case study example presents an AcciNet analysis of the Leeds United scenario.

A task network of elite football club operation is presented in Figure 12.4. The task network was developed from publicly available documents detailing football club operation and the issues associated with the demise of Leeds United. The process undertaken for this case study followed the steps described above. Figure 12.4 shows the high-level tasks required to run a football club at the elite level.

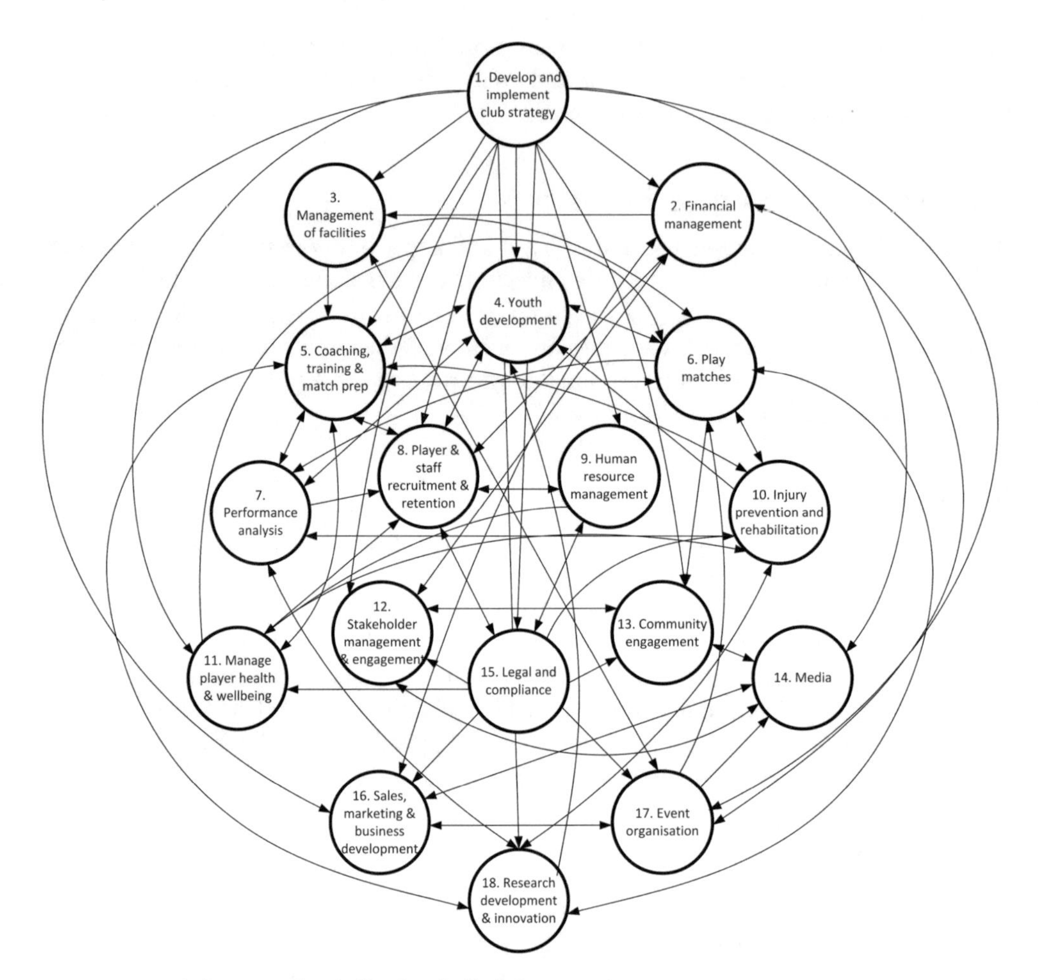

FIGURE 12.4 Task network of elite football club operations.

As shown in Figure 12.4, the task network includes 18 tasks, ranging from the development and implementation of club strategy, financial management, event organisation, player and staff recruitment and retention, playing games, and performance analysis.

An AcciNet showing the contributory factors associated with tasks that interacted to create the Leeds United scenario is presented in Figure 12.5. Within Figure 12.5 the nodes shaded in black represent tasks that were not performed optimally, with the relevant contributory factor code added (e.g., T3 = Task completed inadequately). For example, the task of financial management was undertaken inadequately as the board continued to sanction the purchase of costly players despite the club having significant debt. The corresponding AcciNet table showing a description of the contributory factors is presented in Table 12.5.

Discussion

The AcciNet analysis demonstrates how multiple failures led to the downfall of Leeds United Football Club. As shown in the AcciNet diagram, a central and influential contributory factor was the adoption of a high-risk strategy that involved borrowing

FIGURE 12.5 AcciNet showing contributory factors which led to the downfall of Leeds United.

money on the basis of continual qualification to the Champions League. This is linked to financial mismanagement, which in turn means that once the club did not qualify for the Champions League, they were unable to recruit or keep hold of the talent required to compete at the highest level. As a result, the playing squad began to produce substandard performances, which in turn resulted in relegation. The analysis thus demonstrates the importance of football club strategy and higher-level management in ensuring optimum performance (Salmon et al., 2020). Based on a CLD of English Premier League club, Salmon et al. (2020) found that club strategy and management had the most outgoing connections to other variables, suggesting that it is the most prominent in terms of influence on other variables in the model. The AcciNet analysis provides further support for this, showing how a sub-optimal strategy led to a breakdown in financial management, player recruitment and retention, and ultimately performance. The analysis also suggests that finances act as a natural 'limit to success' in elite football. The limits to success system archetype describe how improved performance and associated successes are slowed down and ultimately ended by natural limits within the system (Kim, 1993; Senge, 1990). In the case of Leeds United, the increasing costs associated

TABLE 12.5 AcciNet output table for Leeds United scenario

Task	*Actor(s)*	*Contributory factor description*	*Contributory factor classification*	*Recommended intervention(s)*
1. Develop and implement club strategy	Chairman Board of Directors	Club developed and implemented high-risk strategy that involved high borrowing on the basis of continual qualification to the Champions League	T3 – Task completed inadequately	– Financial controls and regulation – Undertake independent risk assessment on new club strategy and policy – Improved strategy development process, including independent review and auditing
2. Financial management	Chairman Board of Directors	Club finances were not adequately managed – this included continuing a costly player recruitment strategy despite club having significant debts	T3 – Task completed inadequately	– Financial controls and regulation – Undertake independent risk assessment on new club strategy and policy – Improved strategy development process, including independent review and auditing
6. Play matches	Players Head coach	As the quality of the playing squad declined, performance standards degraded, resulting in relegation, first from the Premier League and subsequently the Championship.	T3 – Task completed inadequately	Maintain quality of playing squad Improved coaching and player development Increased focus on youth development
8. Player and staff recruitment and retention	Chairman Board of Directors Head coach	As a result of large debts the club were forced to sell high-value squad members such as Rio Ferdinand, Robbie Keane, Robbie Fowler, Oliver Dacourt, and Jonathon Woodgate.	T3 – Task completed inadequately	– Financial controls and regulation – Undertake independent risk assessment on new club strategy and policy – Improved strategy development process, including independent review and auditing
13. Community engagement	Players	Leeds players (Jonathon Woodgate and Lee Bowyer) involved in an assault outside a Leeds nightclub in January 2000.	T5 – Inappropriate task	– Player education

with borrowing and maintaining a playing squad of sufficient quality to qualify for the Champions League ultimately led to their downfall.

Recommended Reading

Salmon, P. M., Stanton, N. A., Walker, G. H., Hulme, A., Goode, N., Thompson, J., & Read, G. J. (2022). *Handbook of systems thinking methods*. CRC Press.

References

Chicco, D., & Jurman, G. (2020). The advantages of the Matthews correlation coefficient (MCC) over F1 score and accuracy in binary classification evaluation. *BMC Genomics*, 21(1), 1–13.

Dekker, S. (2011). *Drift into failure: From hunting broken parts to understanding complex systems*. Surrey: Ashgate.

Goode, N., Salmon, P. M., Lenne, M., & Finch, C. (2018). *Translating systems thinking into practice: A guide to developing incident reporting systems*. Boca Raton, FL: CRC Press.

Green, D. M., & Swets, J. A. (1966). *Signal detection theory and psychophysics*. New York: Wiley New York.

Hollnagel, E. (2012). *FRAM, the functional resonance analysis method: Modelling complex socio-technical systems*. Surrey: Ashgate Publishing, Ltd.

Hulme, A., Stanton, N. A., Walker, G. H., Waterson, P., & Salmon, P. M. (2019). What do applications of systems thinking accident analysis methods tell us about accident causation? A systematic review of applications between 1990 and 2018. *Safety Science*, 117, 164–183.

Hulme, A., Stanton, N. A., Walker, G. H., Waterson, P., & Salmon, P. M. (2020). Complexity theory in accident causation: Using AcciMap to identify the systems thinking tenets in 11 catastrophes. *Ergonomics*, 91, 103297.

Hulme, A., McLean, S., Dallat, C., Walker, G. H., Waterson, P., Stanton, N. A., & Salmon, P. M. (2021a). Systems thinking-based risk assessment methods applied to sports performance: A comparison of STPA, EAST-BL, and Net-HARMS in the context of elite women's road cycling. *Applied Ergonomics*, 91, 103297.

Hulme, A., Stanton, N. A., Walker, G. H., Waterson, P., & Salmon, P. M. (2021b). Are accident analysis methods fit for pupose? Testing the criterion-referenced concurrent validity of AcciMap, STAMP-CAST and AcciNet. *Safety Science*, 144, 105454.

Kirwan, B., & Ainsworth, L. K. (1992). *A guide to task analysis: the task analysis working group*. Boca Raton, FL, CRC press.

Klein, G. A., Calderwood, R., & Macgregor, D. (1989). Critical decision method for eliciting knowledge. *IEEE Transactions on Systems, Man, and Cybernetics*, 19(3), 462–472.

Leveson, N. J. (2004). A new accident model for engineering safer systems. *Safety Science*, 42(4), 237–270.

Leveson, N. J., Dulac, N., Marais, K., & Carroll, J. (2009). Moving beyond normal accidents and high reliability organizations: A systems approach to safety in complex systems. *Organization Studies*, 30(2–3), 227–249.

NTSB. (2018). *National Transportation Safety Board Preliminary Report Highway: HWY18MH010*. Retrieved 15 June, Accessed Date, from https://www.ntsb.gov/investigations/AccidentReports/Pages/HWY18MH010-prelim.aspx.

Rasmussen, J. (1997). Risk management in a dynamic society: A modelling problem. *Safety Science*, 27(2–3): 183–213.

Salmon, P. M., McLean, S., Dodd, K., & Stevens, N. (2020). Round and round and up and down we go again: Using causal loop diagrams to model football club performance. In *Human Factors and Ergonomics in Sport* (pp. 255–268). Boca Raton, FL, CRC Press.

Salmon, P. M., Stanton, N. A., Walker, G. H., Hulme, A., Goode, N., Thompson, J., & Read, G. J. (2022). *Handbook of systems thinking methods*. Boca Raton, FL, CRC Press.

Salmon, P. M., Cornelissen, M., Trotter, M. J. (2012). Systems-based accident analysis methods: A comparison of Accimap, HFACS, and STAMP. *Safety Science*, 50(4), 1158–1170.

Salmon, P. M., Hulme, A., Walker, G. H., Waterson, P., Berber, E., & Stanton, N. A. (2020a). The big picture on accident causation: A review, synthesis and meta-analysis of AcciMap studies. *Safety Science*, 126, 104650.

Salmon, P. M., Hulme, A., Walker, G. H., Waterson, P., & Stanton, N. A. (2020b). The Accident Network (AcciNet): A new accident analysis method for describing the interaction between normal performance and failure. *Proceedings of the Human Factors and Ergonomics Society Annual Meeting*. Los Angeles, CA: SAGE Publications Sage CA.

Salmon, P. M., Walker, G. H., Read, G. J. M., Goode, N., & Stanton, N. A. (2017). Fitting methods to paradigms: Are ergonomics methods fit for systems thinking? *Ergonomics*, 60(2), 194–205.

Stanton, N. A. (2006). Hierarchical task analysis: Developments, applications, and extensions. *Applied Ergonomics*, 37(1), 55–79.

Stanton, N. A. (2014). Representing distributed cognition in complex systems: How a submarine returns to periscope depth. *Applied Ergonomics*, 57(3), 403–418.

Stanton, N. A., & Harvey, C. (2017). Beyond human error taxonomies in assessment of risk in sociotechnical systems: A new paradigm with the EAST 'broken-links' approach. *Ergonomics*, 60(2), 221–233.

Stanton, N. A., Salmon, P. M., & Walker, G. H. (2018). *Systems thinking in practice: Applications of the event analysis of systemic teamwork method*. Boca Raton, FL: CRC Press.

Stanton, N. A., Salmon, P. M., Walker, G. H., & Stanton, M. J. (2019). Models and methods for collision analysis: A comparison study based on the Uber collision with a pedestrian. *Safety Science*, 120, 117–128.

Svedung, I., & Rasmussen, J. (2002). Graphic representation of accident scenarios: Mapping system structure and the causation of accidents. *Safety Science*, 40(5), 397–417.

PART 4

Many Model Applications

13

A MANY MODELS APPROACH TO COMPLEX SPORT SYSTEMS ANALYSIS AND DESIGN

This book has introduced ten system thinking-based methods that researchers and practitioners can apply to help understand and respond to complex issues in sport. While each method is useful when applied in isolation, explanatory power is enhanced when using the methods in an integrated manner as part of a many models approach (Salmon & Read, 2019; Salmon et al., 2022). In this final chapter, we demonstrate how an integrated set of systems thinking methods could be applied to tackle one of the most complex and intractable of problems in sport – doping.

Doping is a complex and multifaceted issue that impacts athlete health and wellbeing and compromises the integrity and spirit of sport (Houlihan & Vidar Hanstad, 2019). Further, doping scandals erode the ethical foundation of sport, casting doubt on the legitimacy of achievements and tarnishing the reputation of institutions involved. Financially, the resources allocated to develop more sophisticated testing protocols and enforcement mechanisms represent a substantial economic burden on sporting bodies. Despite the comprehensive anti-doping policies and advanced detection methods in place, the complexity of doping practices, including the emergence of novel substances and masking agents continues to outpace current preventative measures. The dynamic nature of doping strategies complicates the efforts of anti-doping authorities, indicating an imperative for innovative and systemic approaches to assist in doping prevention (McLean et al., 2023).

A many models approach to understanding and preventing doping in sport

Within complexity science, applying a many models approach is advocated when faced with complex and persistent problems (Page, 2016; 2017). This involves using multiple modelling methods to create multiple analyses, or 'models' of the system or issue being investigated, with the aim being to gather insight regarding causal mechanisms, potential interventions, and where possible simulation of likely impacts. By developing a diverse set of models rather than one in isolation, analysts can reach a more detailed understanding,

DOI: 10.4324/9781003259473-17

make better predictions, and ultimately design and implement better interventions (Page, 2016; 2017; Salmon et al., 2022). In the context of the methods presented in this book, it is suggested that greater utility will be achieved by applying a combination of the methods rather than methods in isolation, where possible.

The methods outlined in this book have been categorised into systems analysis and design methods and systemic risk and accident analysis methods. The methods presented in these categories can therefore be used to provide different perspectives on the particular challenge being addressed. Systems analysis and design methods are used to develop models to help understand the structure, composition, and behaviour of a particular system, and to help develop design interventions which aim to optimise system behaviour and performance (Salmon et al, 2022). Systemic risk methods provide a prospective analysis of the vulnerabilities and adverse outcomes that could arise, while accident analysis methods are used to generate in-depth post-hoc analyses of incidents (Table 13.1). By incorporating these approaches using the methods presented in this book, we propose that a many models approach (Figure 13.1) will provide valuable insights into how and why doping occurs and how best it can be prevented. While the remainder of this chapter focuses on doping, the reader is encouraged to think about how a many models approach could be used for other intractable issues in sport, such as injury, performance analysis, technology insertion, and integrity issues such as child safeguarding, corruption, and match-fixing, among others.

As shown in Figure 13.1, the first step involves the use of an appropriate accident analysis method to provide an in-depth analysis of the problem being tackled (Salmon et al., 2022). For example, in Chapter 11, an ActorMap for anti-doping in Australian Rugby is presented, which provides a detailed analysis of the relevant stakeholders tasked with doping prevention. Also in Chapter 11, we presented a case study Accimap's of the contributory factors to adverse incidents and successful performance. Together, these approaches

TABLE 13.1 Summary of the models presented in this book used for the multi-systems-based model approach to preventing doping in sport

Method	*Theoretical underpinning*	*Scope and aims of analysis*
ActorMap AcciMap (Rasmussen, 1997)	Systems theory	To identify the relevant stakeholders tasked with doping prevention, and the contributory factors from across the entire sporting system which are influencing doping in sport.
The systems theoretic accident model and processes (STAMP) (Leverson, 2004)	Control theory and systems theory	To identify the relevant stakeholders tasked with doping prevention, and model the control and feedback mechanisms currently in place to manage doping in Australia.
Work domain analysis (Vicente, 1999)	Ecological psychology	To understand the functional system structure that enabled the RDS.
Hierarchical task analysis (HTA) (Annett et al., 1971)	Control theory	To decompose the behaviours and actions required by athletes seeking to optimise health and performance through supplement use.
System-theoretic process analysis (STPA) (Leverson, 2004)	Control theory	To identify potential control and feedback failures within the Australian anti-doping system.

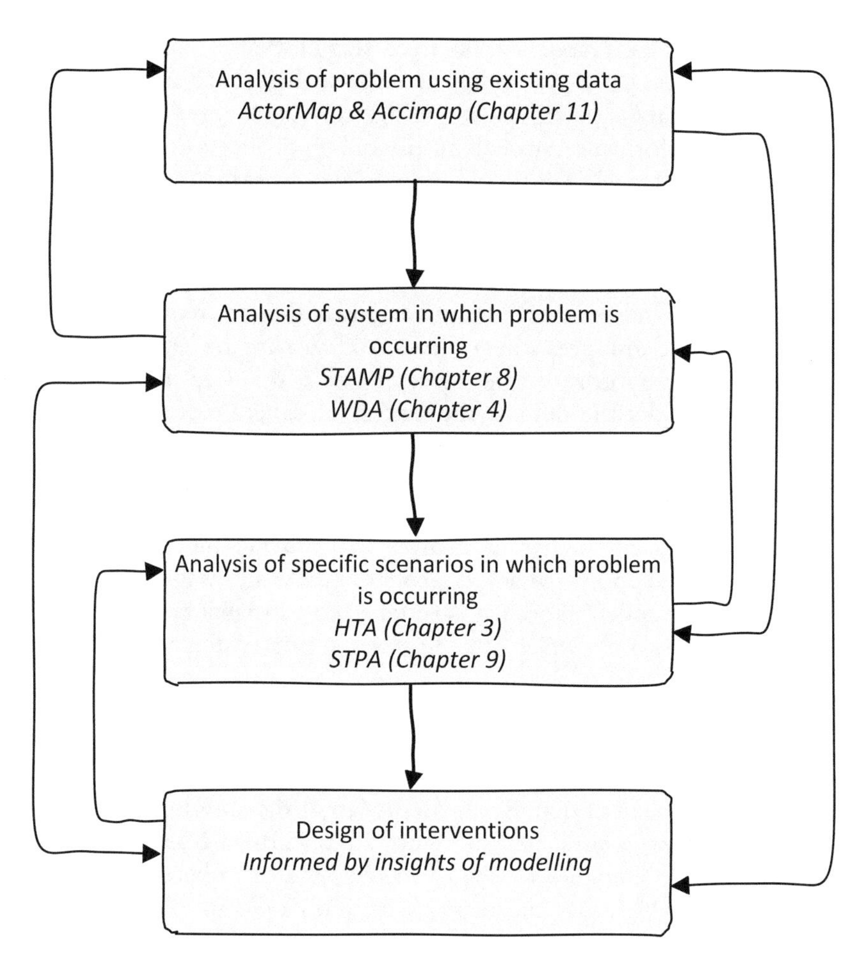

FIGURE 13.1 A many models approach with example methods for use during each step (adapted from Salmon et al., 2022).

can provide detailed insights into the causal mechanisms underpinning the problem of doping in sport as well as the stakeholders who share the responsibility for anti-doping.

The second step of the many models approach involves modelling the system in which the problem is occurring, independent of a specific incident, event, or occurrence. In Chapter 8, a STAMP control structure model of the anti-doping system in Australia was presented. This model presents a detailed description of the control and feedback mechanisms currently in place to manage doping in Australia. This allows us to understand where controls and feedback mechanisms may need strengthening, or where there are opportunities to develop and implement new controls and feedback mechanisms. The control structure also enables us to determine whether certain actors and organisations are overloaded, whether they could do more, or whether new actors are required. In Chapter 4, we introduced Work Domain Analysis (WDA) which is used to understand the functional structure of a given system. In this chapter, we apply WDA to understand the Russian doping scandal, to understand its purposes, measures, tasks, and objects within the system and how they are connected.

Once the system has been modelled, the third step of the many models approach involves the analysis of specific scenarios in which the problem is occurring. In Chapter 3, a Hierarchical Task Analysis (HTA) decomposing the behaviours required by athletes to optimise health and performance through nutritional supplements to avoid unintentional doping was presented. The HTA included tasks related to anti-doping activities as well as tasks around the need for using supplements, and researching of supplements, among others. In Chapter 9, STPA was used to to identify potential control and feedback failures within the Australian anti-doping system. Both analyses give insights into some of the contextual factors that influence doping, the activities that are required for safety and compliance and point to instances where current controls are inadequate or can fail.

At this point the three forms of analysis have provided a deep understanding of the causal mechanisms involved in doping, the functional structure of the anti-doping system, the stakeholders who share responsibility for anti-doping, the controls and feedback mechanisms currently used to prevent doping, potential control and feedback mechanism failures, and the activities and context factors that could lead to doping violations. Taken together this represents a comprehensive analysis that provides multiple insights into the problem of doping and supports the development of interventions to tackle the issue. The final step in the many models approach is to take these insights and use them to drive a participatory design process which aims to develop interventions to resolve or better manage the issue.

Step 1: Analysis of the problem

To understand the problem of doping, an ActorMap of the anti-doping stakeholders in Australian Rugby Union is presented in Figure 13.2. Further, Naughton et al. (2024) conducted a systematic literature review to identify the contributory factors influencing doping in sport, which was subsequently mapped onto the AcciMap framework (Figure 13.3).

The ActorMap demonstrates the diverse set of actors and organisations responsible for anti-doping in Australian Rugby, from an international level (e.g., WADA, IOC), Australian Government (e.g., national and state parliaments), Regulatory bodies (e.g., Sport Integrity Australia, Australian Institute of Sport), teams and organisations (e.g., Rugby Australia, Super Rugby clubs), direct supervisors (e.g., coaches, doctors), the athletes, teammates and opposition, and equipment and environment (e.g., technology, facilities, and doping testing equipment).

Key insights into the causal mechanisms of doping were identified from the AcciMap analysis (Naughton et al, 2024). The findings show that there are contributory factors at each level of the system from international influences and government levels, through to the equipment and substances at the lower levels of the system. The coach and coach-athlete relationship are particular aspects which appear to have a large influence on the doping 'worldview' of athletes. A further finding was that the current deterrence-detection system appears to lack legitimacy to athletes as they perceive the penalties are low relative to the potential rewards on offer for doping-related improvements in performance. Given that the current anti-doping system has had only variable success in doping prevalence, research examining interventions to improve the anti-doping system is recommended. This should include a specific focus on system leverage points which have the potential to elicit large-scale systemic change (Meadows, 2008). The findings

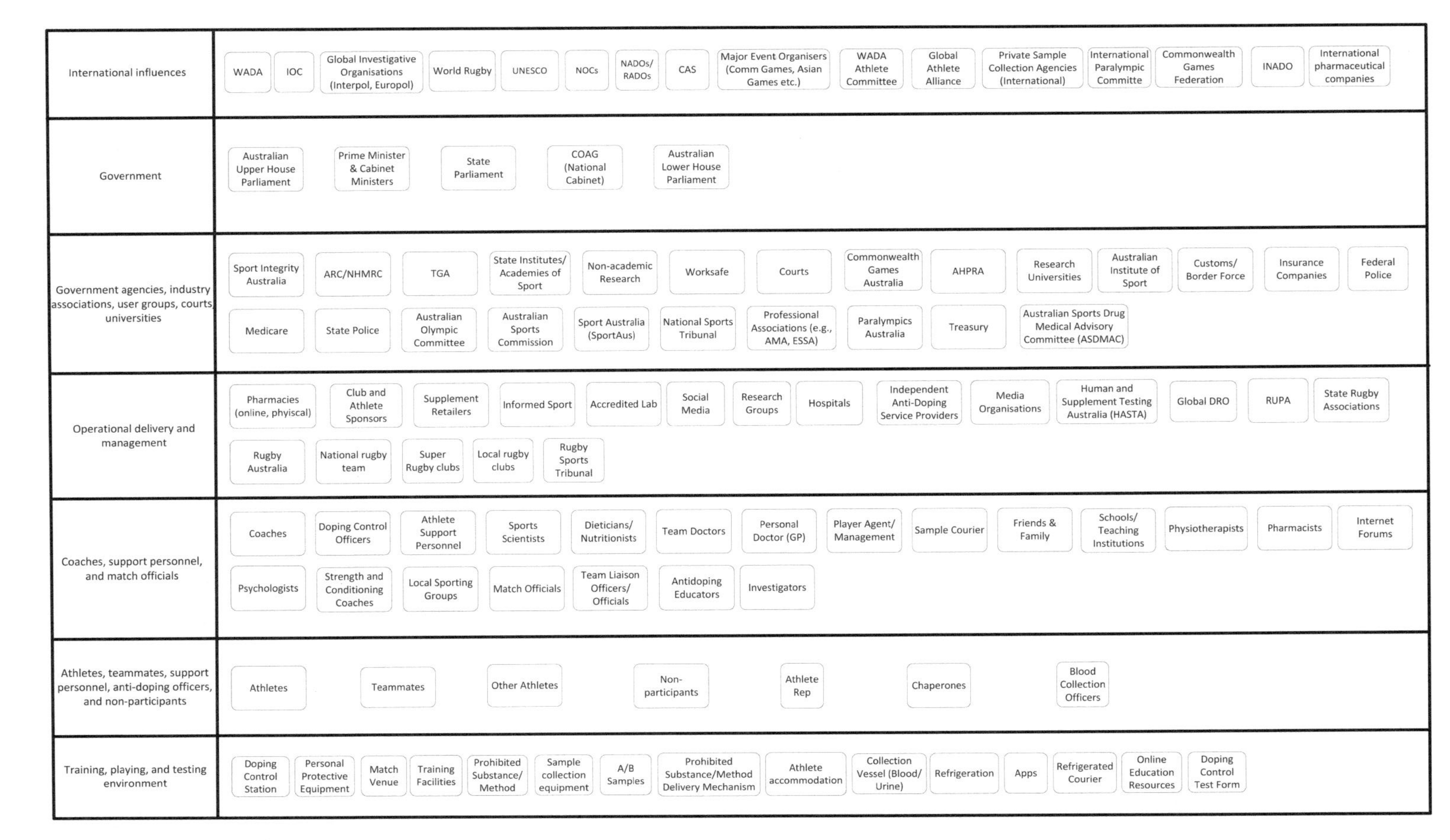

FIGURE 13.2 ActorMap of anti-doping stakeholders in Australian Rugby.

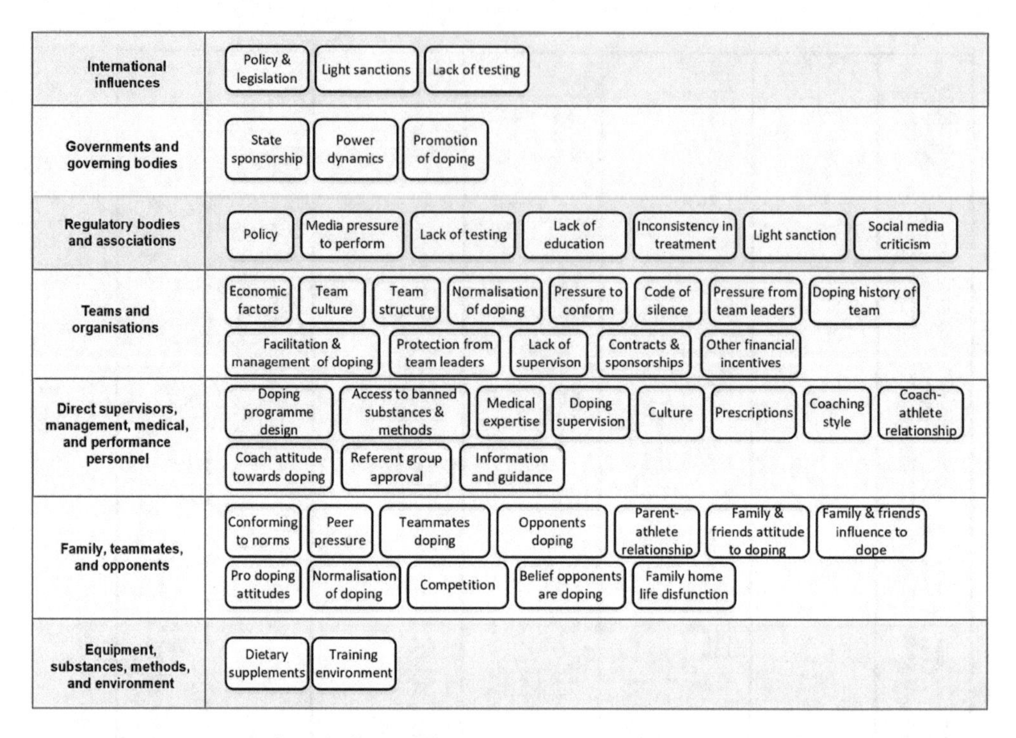

FIGURE 13.3 AcciMap of contributory factors to doping which are identified in the literature (adapted from Naughton et al., 2024).

have implications for athlete health, anti-doping authorities and policymakers as they seek to minimise and, where possible, eliminate doping from sport.

Step 2: Analysis of the system in which the problem is occurring

In Chapter 8, we presented a STAMP control structure of the Australian Rugby anti-doping system to identify what actors share the responsibility for anti-doping, and the control and feedback mechanisms that are currently used to prevent doping (Figure 13.4). In Chapter 4, we presented Work Domain Analysis (WDA), the first phase of Cognitive Work Analysis (CWA). For demonstration purposes of the importance of understanding a system's structure and functioning, in this chapter, we have developed a WDA abstraction hierarchy based on independent reports (McLaren, 2016) to demonstrate and understand the functional system structure that enabled the Russian doping scandal (RDS) (Figure 13.5). The RDS was a state-sponsored doping program that enabled athletes who were doping to avoid detection and win medals at major championships (including Olympics and Winter Olympics). The state-sponsored doping efforts were uncovered through whistle-blowers and, following a series of investigations, ultimately led to the Russian athletes being banned from major international competitions (McLaren, 2016).

The Rugby anti-doping control structure model (Figure 13.4) demonstrates the complexity of the anti-doping system through the specification of numerous actors that share responsibility for anti-doping, and the numerous control and feedback mechanisms. The conceptualisation of the anti-doping system as a complex sociotechnical system (STS)

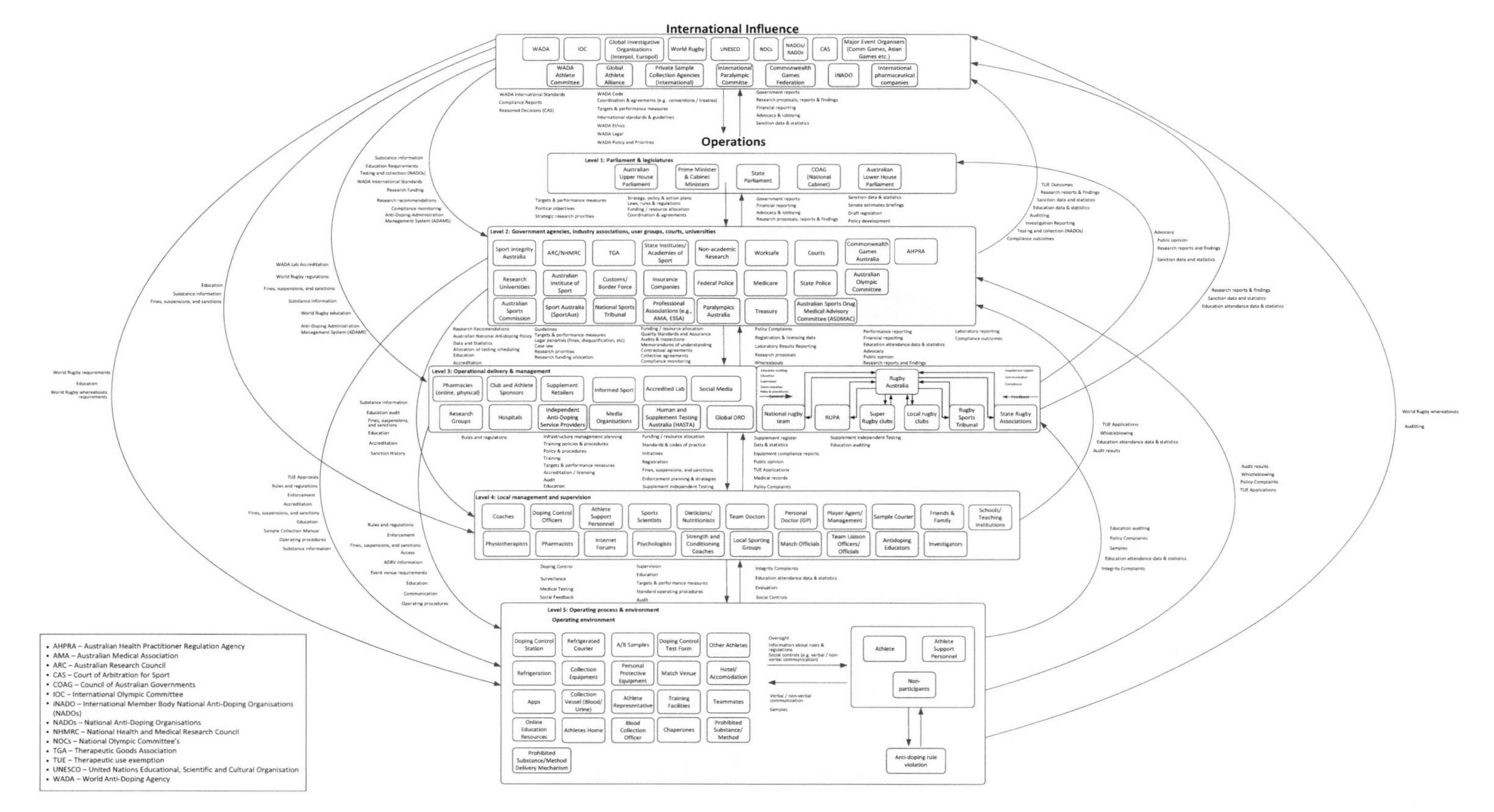

FIGURE 13.4 Australian Rugby Union anti-doping STAMP control structure model.

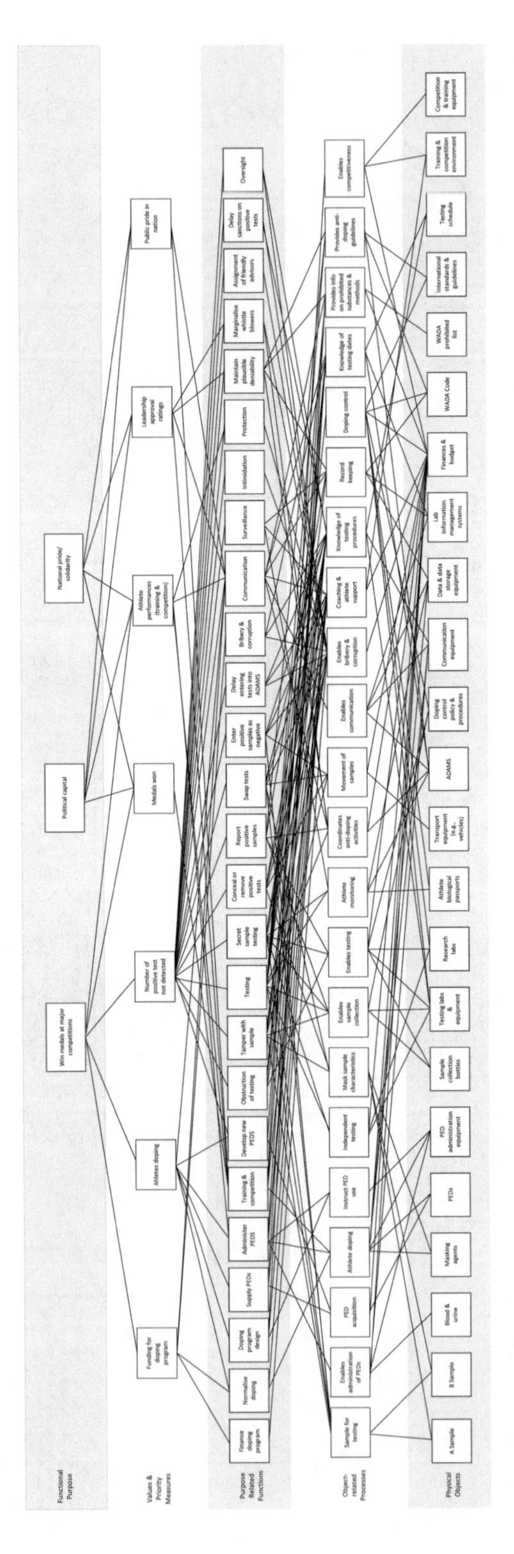

FIGURE 13.5 WDA analysis of the RDS.

supports assertions that doping activities are emergent properties of the broader sports system (Houlihan & Vidar Hanstad, 2019; McLean et al., 2023). The findings suggest moving beyond reducing the problem of doping to a single actor, action, or piece of legislature and consider the role of the broader doping prevention system, its interrelated components, and resulting emergent properties (McLean et al., 2023). The control structure model demonstrated that doping is caused by interacting networks of contributory factors brought about by decisions and actions from actors across the system hierarchy (McLean et al., 2023; Salmon et al., 2020). The numerous controls identified in the model emphasise a profoundly bureaucratic system that is focused on education, detection, deterrence, and enforcement of actors at the lower levels of the system. Further, the control structure identified that education is a key strategy for anti-doping agencies, yet feedback mechanisms including information regarding the effectiveness and reach of the education are missing.

The WDA abstraction hierarchy (Figure 13.5) for the RDS shows the functional structure of the system that enabled the systematic doping program. Through understanding the purposes, measures, tasks, and objects within the system and how they are connected, it is possible to understand the functional structure. Through investigating the means-ends-links in the abstraction hierarchy it was concluded that the Russian doping system was enabled despite the presence of numerous WADA-mandated physical objects including the 'World anti-doping code', the 'WADA prohibited list', 'athlete biological passport', and various testing equipment. This was achieved through multiple purpose-related functions and values which sought to undermine the integrity of testing and sampling, e.g., 'obstruction of testing', 'tampering with collected samples', 'secret sample collections'. Further purpose-related functions carried out to avoid authorities included 'bribery', 'intimidation', 'protection', 'surveillance', 'marginalising whistle-blowers', and 'assigning friendly advisors'. Many of these purpose-related functions were highly connected to nodes at the values and priority level 'number of positive tests' and 'athlete performances', which connected up to the three functional purposes of 'winning medals at major competitions', 'enhancing national pride', and 'increasing political capital'. The application of WDA could be used to develop strategies which can be used in future anti-doping efforts.

Step 3: Analysis of specific scenarios in which the problem is occurring

In Chapter 3, to better understand the occurrence of inadvertent doping through nutritional supplement use we presented a HTA decomposing the behaviours and cognitive processes undertaken by athletes attempting to optimise health and performance through supplements (Figure 13.6). In Chapter 9, we presented an STPA to identify potential control and feedback failures within the Australian anti-doping system with the aim of identifying areas to strengthen controls and feedback mechanisms or develop new ones (Figure 13.4).

The HTA demonstrates the numerous sub-goals and operations involved when athletes seek to use nutritional and dietary supplements to improve health and performance. As such, it is understandable that athletes are susceptible to inadvertent doping through supplement use. Decomposing this complexity through a HTA can help to identify potential areas for improvements in the process. For example, the HTA highlights that within the current system there are limited opportunities for intervention from outside

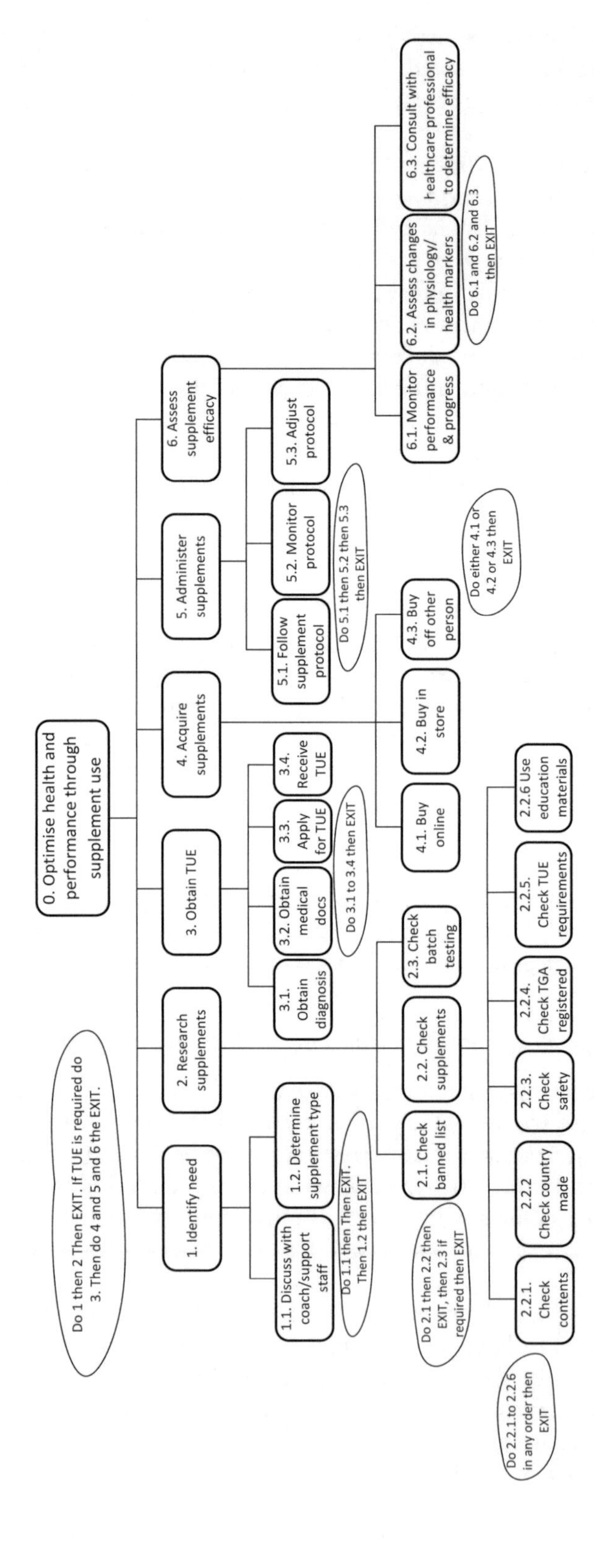

FIGURE 13.6 A HTA decomposition of the behaviours and cognitive processes athletes undertake when using supplements for health and performance.

of the athlete. Only the discussion with coaches and support personnel, and obtaining a TUE represent potential intervention points to assist with athlete decision-making and actions. However, athletes are solely responsible for driving these steps, and are not required to complete them prior to supplement use.

The STPA analysis provides useful insights about how the anti-doping system could potentially fail, as well as risk descriptions that are relevant to numerous stakeholders within the anti-doping system. The STPA revealed a total of 712 potential failures, including 428 control failures and 284 feedback failures. While some of these potential failures may be currently well managed through existing control and feedback mechanisms, some may not be. Consequently, the list of identified potential failures provides a useful database for review when considering the effectiveness of current control and feedback mechanisms and the development of new interventions. It is recommended that sport system stakeholders review the STPA risks and their current control and feedback mechanisms to determine if further effort around the development or strengthening of control and feedback mechanisms is required.

For the controls, while almost a third of the potential failures relate to the operating process and environment level and the controls imposed on athletes (e.g., testing, education, enforcement), the majority relate to potential failures of the controls imposed on other actors. For example, 76 potential failures were identified for the controls imposed by actors at the *government agencies* level on actors at the *operational delivery and management* level. These included potential failures around the quality and implementation of the national anti-doping policy, compliance monitoring, resource allocation, quality standards and assurance, and accreditation. Similarly, 40 potential failures were identified for the controls imposed by international actors on actors at the parliament and legislature level. For example, these included potential failures relating to the content and implementation of the WADA code, ethics guidelines, and policies and procedures. These findings demonstrate the critical need to consider risk assessment and management for all stakeholders within the anti-doping system as opposed to athlete only controls. While testing currently provides a control for athlete doping, there may be other opportunities to strengthen the anti-doping system. For example, enhancing controls such as 'funding and resource allocation' and 'allocation of testing schedule' will in turn enhance testing outcomes, as more athletes can be tested in an appropriate manner, and testing protocols can keep pace with advances in performance-enhancing drugs.

A summary of the insights obtained through the application of a many models approach for doping in sport is presented in Table 13.2. It is suggested that organisations include another column in Table 13.2 to include interventions they have designed to respond to the issues that have been identified. See Step 4 for guidance on intervention design.

Step 4: Designing interventions based on model insights

The modelling insights presented in Table 13.2 provide guidance of where to intervene with mitigation strategies for doping in sport. Although we have not included any chapters on designing interventions in the current book, the following section will provide some guidance for developing systemic interventions.

TABLE 13.2 A summary of the insights of the many models approach to doping in sport

Insight	*Method*
Anti-doping systems are complex and comprise numerous actors and organisations that share the responsibility for anti-doping.	STAMP ActorMap
Doping in sport is an emergent property of the broader complex sports system though the interactions of system components.	STAMP WDA AcciMap
Anti-doping rule violations are created by multiple contributory factors rather than the single action or decision made by an athlete.	AcciMap STAMP
While there are contributory factors to doping at each level of sport systems from international influences and government levels through to the equipment and substances at the lower levels of the system, the majority of research on the contributory factors to doping is focused on the lower levels of the system, e.g., athletes and support personnel.	AcciMap
The anti-doping system contains numerous controls that are focused on education, detection, deterrence, and enforcement of actors at the lower levels of the system.	STAMP
The anti-doping system controls are reactive to anti-doping rule violation incidents, rather than comprising proactive controls to prevent doping instances in the first place.	STAMP
Anti-doping systems lack appropriate feedback mechanisms to ensure policymakers are informed regarding anti-doping, especially the reach and effectiveness of education efforts.	STAMP STPA
The functional structure of anti-doping systems potentially enables doping to occur, despite internationally mandated codes, rules, regulations, and policy.	WDA
Athletes are expected to be self-compliant with the anti-doping regulations around the use of supplements to optimise health and performance.	HTA
Athletes seeking to use supplements to optimise health and performance face a complex set of goals, sub-goals, and operations in order to avoid inadvertent doping with little help outside of education materials.	HTA
Existing doping incident reporting systems are inadequate as they do not support the collection of data on contributory factors across multiple levels of the anti-doping system.	STAMP STPA

One approach to developing systemic interventions is the Sociotechnical Systems Design Toolkit (STS-DT) (Read et al., 2018). The STS-DT is underpinned by a socio-technical systems theory (STS) and supports the design of complex systems that exhibit adaptive capacity and align with core STS values (see Table 13.3). The STS-DT provides a framework for understanding the social and technical aspects of a system and guides the design process to develop interventions that consider both dimensions. By analysing the social context, stakeholder needs, and system requirements, the STS-DT helps identify areas where interventions may be necessary. Interventions designed using the toolkit can take various forms depending on the specific context and goals. They can range from technological changes to organisational restructuring, policy development, or training programs. The STS-DT helps designers and practitioners consider the potential impacts of these interventions on both the technical and social aspects of the system. The STS-DT also emphasises the importance of involving stakeholders in the design process. By engaging end-users, employees, and other relevant stakeholders, interventions can be co-designed and co-created to ensure their relevance, effectiveness, and acceptance within the STS.

TABLE 13.3 The STS design values and examples from anti-doping in sport

Design value	*Anti-doping example*
1 Humans should be treated as assets rather than as unpredictable, error-prone, and the cause of problems in otherwise well-designed systems.	Involving athletes in the development of anti-doping policies and educational programs acknowledges their valuable insights and promotes a collaborative approach to clean sport.
2 Technology should be used as a tool to assist humans, rather than being seen as an end in its own right.	We should not fall into the trap of viewing technologies, such as blood and urine analysis tools, biological marker tracking, and even genetic testing as infallible solutions to eradicate doping independently, they are utilised to complement and support the broader anti-doping efforts, which include education, ethical training, and athlete welfare programs.
3 Designs should focus on promoting quality of life, rather than creating strict work requirements (e.g., lack of flexibility around working hours and breaks), poor work design (e.g., repetitive tasks, lack of task rotation), and unachievable expectations. Quality work should be challenging and incorporate variety, should include scope for decision-making and choice and facilitate ongoing learning, should incorporate social support and recognition, and should have social relevance to life outside work (Cherns, 1976, 1987).	Scheduling of tests in a way that minimises disruption to athletes' training, rest, and personal life respects their need for balance between their sporting and their personal commitments.
4 Designs should respect individual differences in the needs and preferences of the various end-users.	Anti-doping agencies recognise that athletes come from diverse backgrounds, with varying levels of exposure to and understanding of doping issues. Instead of a one-size-fits-all approach to anti-doping education and policies, programs are tailored to address the specific needs, including cultural contexts, learning preferences of different athletes, age and experience, and be sport specific.
5 Designers should consider all stakeholders, including the impacts of choices they make on various stakeholders.	Consideration for stakeholders' perspectives including athletes, but also coaching staff, performance staff, administrators, sponsors, testing staff, fans, law enforcement, and governance.

Discussion

While each of the methods described in this book is useful in their own right, for highly complex problems in sport, such as doping, the application of a many models approach can provide a far more detailed, multi-perspective, and nuanced analysis. This chapter has demonstrated that the many models approach provides a powerful framework for understanding doping occurrence and anti-doping processes across different contexts e.g. unintentional doping through supplement use to state sponsored doping scandals. The findings indicate that it is beneficial to apply multiple systems thinking methods to assess the same problem, with the five analyses producing a series of complimentary insights spanning all levels of the anti-doping systems. Notably, some insights were unique to an individual method, some were repeated across analyses, and none of them appear to be incongruent or at odds with each other (Salmon et al., 2022).

A useful feature of the many models approach is that it permits a progression of work from analysis of the problem and system through to the design of potential solutions (Salmon & Read, 2019). The specific combination of methods to use will depend on the project constraints and scope, however, if the project scope allows it, it is recommended that all four steps described in Figure 13.1 be employed. Although, the application of a many models approach will require substantial resources, the detailed outputs represent a beneficial effort to gain ratio. If project constraints and scope do not permit application of all four steps, incorporating a smaller selection of the steps and methods will still prove useful. For example, one less resource intensive approach might involve using AcciMap to analyse a particular incident, followed by the use of WDA to analyse the system, specific behaviours, and then to design appropriate solutions (Salmon & Read, 2019).

Recommended Reading

Salmon, P. M., & Read, G. J. M. (2019). Many-model thinking in systems ergonomics: A case study in road safety. *Ergonomics*, 62(5), 612–628.

Salmon, P. M., Stanton, N. A., Walker, G. H., Hulme, A., Goode, N., Thompson, J., & Read, G. J. (2022). *Handbook of Systems Thinking Methods*. CRC Press.

Read, G. J., Salmon, P. M., Goode, N., & Lenné, M. G. (2018). A sociotechnical design toolkit for bridging the gap between systems-based analyses and system design. *Human Factors Ergonomics in Manufacturing Service Industries*, 28(6), 327–341.

References

Clegg, C. W. (2000). Sociotechnical principles for system design. *Applied Ergonomics*, 31, 463–477. doi: 10.1016/s0003–6870(00)00009-0

Houlihan, B., & Vidar Hanstad, D. (2019). The effectiveness of the World Anti-Doping Agency: Developing a framework for analysis. *International Journal of Sport Policy and Politics*, 11(2), 203–217.

McLaren, R. H. (2016). McLaren Report WADA.

McLean, S., Rath, D., Lethlean, S., Hornsby, M., Gallagher, J., Anderson, D., & Salmon, P. M. (2021). With crisis comes opportunity: Redesigning performance departments of elite sports clubs for life after a global pandemic. *Frontiers in Psychology*, 11, 588959.

McLean, S., Naughton, M., Kerhervé, H., & Salmon, P. M. (2023). From Anti-doping-I to Anti-doping-II: Toward a paradigm shift for doping prevention in sport. *International Journal of Drug Policy*, 115, 104019.

Meadows, D. H. (2008). *Thinking in systems: A primer*. Chelsea Green Publishing. White River Junction, Vermont

Naughton, M., Salmon, P. M., Kerherve, H., & McLean, S. (2024). Applying a systems thinking lens to anti-doping: A systematic review identifying the contributory factors to doping in sport. *Journal of Sports Sciences*. 1–15.

Page, S. (2016). Many model thinking. IEE, Winter Simulation Conference (WSC), 11th – 14th December 2016, Washington, DC.

Page, S. (2017). Many model thinking. Keynote address, Computational Social Sciences of the Americas Annual Conference, October 19–22, Santa Fe.
Read, G. J., Salmon, P. M., Goode, N., & Lenné, M. G. (2018). A sociotechnical design toolkit for bridging the gap between systems-based analyses and system design. *Human Factors Ergonomics in Manufacturing Service Industries*, 28(6), 327–341.
Salmon, P. M., Hulme, A., Walker, G. H., Waterson, P., Berber, E., & Stanton, N. A. (2020). The big picture on accident causation: A review, synthesis and meta-analysis of AcciMap studies. *Safety Science*, 126, 104650.
Salmon, P. M., & Read, G. J. M. (2019). Many-model thinking in systems ergonomics: A case study in road safety. *Ergonomics*, 62(5), 612–628.
Salmon, P. M., Stanton, N. A., Walker, G. H., Hulme, A., Goode, N., Thompson, J., & Read, G. J. (2022). *Handbook of Systems Thinking Methods*. Boca Raton, FL, CRC Press.

INDEX

Note: **Bold** page numbers refer to tables and *italic* page numbers refer to figures.

For Product Safety Concerns and Information please contact our
EU representative GPSR@taylorandfrancis.com Taylor & Francis
Verlag GmbH, Kaufingerstraße 24, 80331 München, Germany